Simulation

Simulation
A Modeler's Approach

JAMES R. THOMPSON

Rice University
Houston, Texas

A Wiley-Interscience Publication
JOHN WILEY & SONS, INC.
New York / Chichester / Weinheim / Brisbane / Singapore / Toronto

Library of Congress Cataloging-in-Publication Data:

Thompson, James R. (James Robert), 1938–
    Simulation : a modeler's approach / James R. Thompson.
        p.   cm. — (Wiley series in probability and statistics.
    Applied probability and statistics)
    "A Wiley-Interscience publication."
    Includes bibliographical references and index.
    ISBN 0-471-25184-4 (alk. paper)
    1. Experimental design.   2. Mathematical models.   3. Mathematical statistics.   I. Title.   II. Series.
    QA279.T494   1999
    003—dc21   99-33022

Printed in the United States of America

10 9 8 7 6 5 4 3 2

To my mother, Mary Haskins Thompson

Contents

Preface

Half a century ago, John von Neumann created the digital computer as a device for carrying out simulations. Although the computer revolution has advanced to the point where 500 MHz boxes are available for under $2000, the simulation revolution is still in its infancy.

The situation where we can produce a probability profile of an investment strategy based on, say, 10,000 simulations with inputs the parameters of that strategy is still more talked about than achieved. The same can be said for a stochastic analysis of a political/military strategy in the Balkans. Stochastic models for a proposed new transportation system in a large city are still not well developed. Cancer profiles for individual patients based on their immune systems and the likely presence of metastases at the time of diagnosis are not readily available. Stochastic profiles of an epidemic vectored into the United States by a hostile power, based on the modality of its introduction and transmission pathways, are undeveloped. Particularly in simulation, software vision has lagged hardware by decades. A major function of this book is to indicate possibilities for synergy between data, models and the digital computer.

What is simulation? Presumably a simple question, but the scientific community is far from a consensus as to the answer. A government administrator might decide to "simulate" the national effect of a voucher system by taking a single school district and implementing a voucher system there. To a geologist, a simulation might be a three-dimensional differential-integral equation dynamic (hence, four-dimensional) model of implementation of tertiary recovery for an oil field. To a numerical analyst, a simulation might be an approximation-theoretic pointwise function evaluator of the geologist's dynamic model. To a combat theorist, a simulation might consist of use of the Lanchester equations to conjecture as to the result of a battle under varying conditions. To a nonparametric bootstrapper, simulation might consist in resampling to obtain the 95% confidence interval of the correlation coefficient between two variables.

While all of the above may be legitimate definitions of *simulation*, we shall concentrate on the notion of a simulation being the generation of pseudodata on the basis of a model, a database, or the use of a model in the light of a database. Some refer to this as *stochastic simulation*, since such pseudodata tends to change from run to run.

A model? What is that? Again, the consensus does not exist. We shall take a model to be a mathematical summary of our best guess as to what is going on in a part of the real world. We should not regard a model as reality, or even as a stochastic perturbation of reality, only as our current best guess as to a portion of reality. We should not be surprised if today's model is considered rather poor ten years hence. Some people quite mistakenly try to build the biggest model they can. It is not unusual for models of really big systems (such as the world) to be completely artificial. (Club of Rome models come to mind.)

In attempting to come up with one big all-encompassing theory of everything, one loses a great deal. Compartmentalization is clearly one way out of the morass. We can try for a theory that works under very specific conditions. A reasonable way to proceed. But like all good ideas, we might carry it to an extreme. Billy takes a test while carrying a good luck charm in his pocket and makes 100. He takes another test without the charm and scores 50. A third test with the charm yields another 100. Inference: the good luck charm produced better scores for Billy than he would otherwise have made, at least on the days when the tests were taken. An extreme case of nominalist logic, but the point is clear enough. *Ad hoc'ery* leads us very quickly to magic rather than to science. Pre-Socratic modalities of thought emerge, and we in trouble.

It is a hallmark of Western thinking that events indexed on time and faithfully recorded give us a database on which inferences might be made. On the other hand, the postmodernist view holds that the recorder of the events creates, more or less arbitrarily, his or her own history, that there is no reality behind the recording. The recording is the history. The recorder has created his or her own reality. If databases are just a reflection of the prejudices of the recorder, the modeler simply concatenates his or her prejudices with those of the creator of the database to give us, well, nothing very useful.

Models are generally oversimplifications, at best. Perhaps it would be better to get the modeler out of the loop. Among those who would like to free us from the bondage of models is University of Southern California Professor Bart Kosko, a recognized leader in neural networks and fuzzy thinking. In *Fuzzy Thinking* , he writes:

> ...linear systems are the toy problems of science and yet most scientists treat real systems as if they were linear. That's because we know so little math and our brains are so small and we guess so poorly at the cold gray unknown nonlinear world out there. ....Fuzzy systems let us guess at the nonlinear world and yet do not make us write down a math model of the world. .... The technical term for it is model-free estimation or approximation. ...
>
> The key is no math model. Model-free estimation. Model

freedom. If you have a math model, fine. But where do you find one? You find good math models only in textbooks and classrooms. They are toy answers to toy problems. The real world pays no attention to most of them. So how do you know how well your math guess fits Nature's process? You don't and you never can. You have to ask God or No God or Nature and no one knows how to do that. Short of that, we guess and test and guess again. Scientists often respect math more than truth and they do not mention that a math guess is no less a guess than a guess in everyday language. At least a word guess does not claim to be more than a guess and gains nothing from the fact that we have stated it in words. A math guess has more dignity the less math you know. Most math guesses are contrived and brittle and change in big ways if you change only a small value in them. Man has walked Earth for at least a million years and has just started to think in math and is not good at it. Fuzzy systems let us model systems in words.

We may well agree with the basic modeling problem mentioned by Kosko, although not at all with his "solution" of it. It is true that in modeling systems mathematically, people tend to be oriented toward models with whose mathematical complexities they can cope. For example, we can postulate that a tumor will grow, moment by moment, in proportion to its mass. And a further postulate can be made that the probability of a metastasis being generated by a tumor in a short period of time is proportional to the mass of the tumor. Another postulate can be made that the probability that a tumor will be discovered in an instant of time is proportional to the mass of the tumor. These postulates represent quite a simple-sounding model for the progression of cancer in a patient. But when one starts looking at the times of discovery of primary and secondary tumors and using this information to estimate the parameters of the simple-sounding model, it is discovered that getting anything like a likelihood function (a general first step in classical parameter estimation) is a hopelessly complicated business. The reason for the problem is that the axioms are made in a forward temporal direction, whereas the likelihood function is computed looking backward in time to the possible causes of generation of particular tumors. Such complexities have caused most biostatisticians to work with linear aggregate models, such as the survival times of patients taking drug A as opposed to drug B. Such analyses have not worked very well, and it is the failure of such simpleminded linearizations, in part, which have made the War on Cancer a series of losing engagements.

But to use a fuzzy system or a neural net as a way out of the linear oversimplification is to replace fiction with magic. We really need to know how cancer grows and spreads. Some glorified smoothing interpolator is unlikely to get us out of the soup. Later, we shall show how simulation

allows us to perform parameter estimation without false linear simplifica-
tions and without an uninformative hodgepodge of neural networks. The
SIMEST paradigm will enable us to use postulated models in such a way
that backwards-in-time mathematical operations can be eliminated in favor
of a large number of forward simulations.

Nearly seven hundred years ago, that most famous of nominalists, William
of Occam, gave a pronunciamento which can be viewed as a prototype of
that of Kosko:

> Nonetheless, one should know that there are two kinds of
> universal. One kind is naturally universal, in that it is evi-
> dently a sign naturally predictable of many things, in a fashion
> analogous to the way smoke naturally signifies fire,.... Such a
> universal is nothing except a notion of the mind, and so no sub-
> stance outside the mind nor any accident outside the mind is
> such a universal....The other kind of universal is so by estab-
> lished convention. In this way a word that is produced, which
> is really one quality, is a universal, because clearly it is a sign es-
> tablished by convention for the signifying of many things. Hence
> just as a word is said to be common, even so it can be said to
> be universal; but this does not obtain from the nature of the
> thing, but only from agreed upon convention.[1]

Nominalism is at odds with the realist (Aristotelian) view of science as
an evolutionary search for better and better descriptions of objective reality. One current fashion in the history of science is to look for fundamental
change points in the dominant scientific paradigm. These change points are
essentially a political phenomenon. This was the view of the late Thomas
Kuhn. For example, the Newtonian relationship between force and momen-
tum is given by

$$F = \frac{d}{dt}(mv), \tag{1}$$

which is normally written as

$$F = ma, \tag{2}$$

where F is force, m is mass, v is velocity and a is acceleration. But following the discovery of Einstein that mass changes as the speed of the
object increases, we are to view Newton's representation as hopelessly out
of date. We reject Newtonianism in favor of Einsteinism and go on about
our business anxiously awaiting the advent of the new evangel which will
trash Einstein.

In this book, we take the more classical notion that Einstein improved
Newton's model rather than making it fit for the dustbin. Consider that

---

[1] *The Sum of All Logic* translation by Philtheus Boehner, O.F.M.

Einstein has

$$m = \frac{m_0}{\sqrt{1 - v^2/c^2}} \qquad (3)$$

where c is the speed of light. We can then take this expression for m and substitute it back in (1) to give

$$F = \frac{d}{dt}\left(\frac{m_0}{\sqrt{1 - v^2/c^2}}v\right). \qquad (4)$$

Beyond that, the Newtonian model gives essentially the same results as that of Einstein for bodies not moving at light speed. Anybody who believes that Newtonian physics has outlived its usefulness has not examined the way mechanics is taught by contemporary departments of physics and mechanical engineering. So we shall take the view of evolution rather than revolution in model progression. It is the view of the author that simulation is simply a computer-facilitated implementation of the model-building process (a.k.a the scientific method).

One might well ask why the necessity for this philosophical digression. Let us just start simulating already. Fair enough. But simulate what? And to what purpose? A great deal of the literature in simulation is oriented to a kind of idealized mathematical formalism. For example, there are many hundreds of papers written on the subject of the generation of random numbers. And very important many of them are. But this can lead us to a kind of formalist copout. If we dwell excessively on an algorithm that yields a string of random numbers which satisfy some kind of arbitrary set of desiderata, we can get lost in the comfortable realm of mathematical theorem statement and proof. All very tempting, but we shall not travel very far down that road. This is a statistics book, and statisticians should be concerned with a reality beyond data-free formalism. That is why statistics is not simply a subset of mathematics.

Our major interest in this book will be using simulation as a computational aid in dealing with and creating models of reality. We will spend a bit of time in going through the philosophy of quasirandom number generators and we will go through some of the old Monte Carlo utilization of simulation in, for example, the approximation of definite integrals. But our main goal will be to use simulation as an integral part of the interaction between data and models which are approximations to the real systems that generated them.

A goodly amount of time will be employed in resampling procedures where we use resampling from a data set (or, in the case of SIMDAT, from the nonparametric density estimator of the density based on the data set) to test some hypothesis and/or obtain some notion of the variability of the data. But much more important will be the use of simulation as an integral part of the modeling process itself. As an example of the latter, let us consider a couple of "toy problems."

We recall the quiz show with master of ceremonies Monty Hall. Three doors were given, say A, B, and C. Behind one of these was a big prize. Behind the others, nothing splendid at all. The contestant would choose one of the doors, say A, and then the MC would tell him one of the other doors, say C, behind which the splendid prize did not exist. Then the contestant was given the option of sticking with his original choice A or switching to B. The quiz show continued for some years in this manner with contestants going both ways, so it is clear that the general consensus was that there was no systematic preference for either strategy. Let us go through a standard Bayesian argument to support this view.

Let us compute the probability of winning of the contestant who "stands pat," i.e., he chose A originally; he learned that the prize was not behind door C, yet he decided to stick with his original choice of door A. Let us compute the probability that he will win given that C is not the prize door.

$$
\begin{aligned}
P(A|C^c) &= \frac{P(A \cap C^c)}{P(C^c)} \\
&= \frac{P(A)P(C^c|A)}{P(C^c)} \\
&= \frac{(1/3)(1)}{2/3} \\
&= .5.
\end{aligned}
$$

The reasoning seems to be correct. The prior probability here is $P(A) = 1/3$. If A is the prize door, the chance that C is not the prize door is 1 [i.e., $P(C^c|A) = 1$]. Finally, the prior probability that C will not be the prize door is 2/3 [i.e., $P(C^c) = 2/3$]. Furthermore, once we have been told that C is not the prize door, then $P(A|C^c) + P(B|C^c) = 1$, so $P(B|C^c)$ must equal .50 as well. A formal argument would seem to support the popular wisdom that it makes no difference, over the long haul, whether contestants stand pat or switch to B.

But, so the story goes, somebody went back over the records of the contestants and found that those who switched, on the average, did better than those who stood pat. In fact, the switchers seemed to win about two thirds of the time. Could this be due to the laws of probability, or was something else afoot? Here is a case where the simulations consisted not of computer simulations, but actual implementations of the game.

To help us out, let us write a simple simulation program.

```
Set counter WA equal to zero
Set counter WSwitch equal to zero
Repeat 10,000 times
Generate U, a uniform random number between 0 and 1
If U is greater than .33333, go to *
Let WA = WA+1
```

Go to **
* WSwitch = WSwitch + 1
** Continue
WA = WA/10000
WSwitch = WSwitch/10000
Print WA and WSwitch
End

The argument here is straightforward. We can associate a random number less than .33333 with a win for A. If the prize is behind door A, the standpat strategy always produces a win. On the other hand, a number greater than .33333 will be associated with a win for B or C. The MC will tell us which of these two doors is not the prize door, so if we switch to the other, we always win. A simulation of 10,000 trials gave a .3328 probability of winning by the standpat strategy, as opposed to a .6672 probability of winning by the switch strategy.

Indeed, once one writes down this simulation, the problem is essentially solved without any simulations. It simply becomes transparent when we flowchart the program. But then, since we do not accept postmodernist and fuzziest notions of the possibility of logical inconsistency, we must needs see where we went wrong (and where the man in the street must have empirically gone wrong). Our writing down of Bayes' theorem is correct. The problem comes in the evaluation of $P(C^c|A)$. We should not interpret this to mean that C is not the prize door when A is the prize door. Rather, it is the probability that the MC will tell us that of the two non-A doors, C is not the prize door. He must pick one of the two B and C with equal probability. Hence $P(C^c|A)$ equals 1/2, rather than unity. Making this correction, we get the probability of winning using the standpat algorithm to be 1/3, as it should be.

Next, let us relate the Monty Hall problem to one of antiquity. Below, we consider one of the many "prisoner's dilemma" problems. A prisoner is one of three condemned to be beheaded on the morrow. But the Sultan, in his mercy, has decided to pardon one of the prisoners. Prisoner A is a mathematician. He wishes to improve his chance of not getting the chop. He knows that the chief jailer knows who is to be spared but has been warned by the Sultan that if he tells any of the prisoners who is to be spared, then he, the jailer, will be disemboweled. The mathematician calls the jailer aside and offers him 100 drachmas to tell him, not the name of the fortunate, but the name of one of those, other than, possibly, himself, who is to be executed. The jailer agrees and tells A that C is one of the condemned. A heaves a sigh of relief, since he now believes that his probability of being spared has increased from one-third to two-thirds. Let us note, however, that A actually is in the position of the standpat player in the Monte Hall example. His probability of survival is actually one-third. On the other hand, if B happens to overhear the exchange between A and his jailer, he does have some reason for relative optimism, since he stands in the position of the switch player in the Monty Hall example. It is, naturally, an easy matter to write down a simulation program for the prisoner's dilemma situation, but the analogue between the two situations is actually an isomorphism (i.e., the problems are the same precisely.

Let us note, here, the fact that the construction of a simulation is, clearly, a kind of modeling process. It will generally cause us to analogize a temporal process, since computer programs consist of instructions which take place in a sequence. A number of the differential and integral equations of physics were natural summarizing models for the precomputer age. Typically, the closed-form solution for fixed parameter values is not available. We must be satisfied with pointwise approximations to the value of the dependent variable vector y in terms of the independent variable vector x. It turns out that in a large number of cases we can approximately carry out this approximation by a simulation, which frequently is based on the microaxioms that gave rise to the differential-integral equation summary in the first place.

Of greater interest still is the situation where we have the postulates for our model (and hence, in principle, the model itself) and a database and wish to estimate the underlying parameters. According to classical paradigms, in order to estimate these parameters, we must obtain something like a likelihood function. But this is generally a hopelessly complicated task. The SIMEST paradigm, which we examine, allows us to go directly from the postulates and the data to the estimation process itself. This is achieved stepwise by creating a large class of pseudodata predicated on the assumption of a particular (vector) parameter value. By comparing the pseudodata with the actual data, we have a natural means of moving to a good estimate of the characterizing parameter. This is a temporally forward estimation procedure, as opposed to the classical estimation strategies which look backwards in time from the data points.

Another use of simulation will be in the realm of scenario analysis. We shall, for example, examine some of the current models for movement of stock and derivative prices and analyze some pricing strategies in the light of changes made in these models. This is a speculative use of simulation. We are not using data intimately. Rather, we wish to ask "what if?" questions and use simulation to give some clue as to feasible answers.

As we have noted, there are many who, discouraged by the results of the use of bad models, would like to dispense with models altogether. And simulation can frequently be put to good use in dealing with model-free analyses. The basic problem of model-free analysis is that it can work well when one is interpolating within a database, but it generally decays rapidly when we start extrapolating. To make matters even more difficult, for data of high dimensionality, even interpolation within the convex hull of the database is, in fact, a problem of extrapolation, since the data will generally be distributed in clusterlike clumps separated by substantial empty space.

Simulation is also used by those who are happy to assume the correctness of a model and get on with the business, say, of obtaining pointwise approximations of a dependent variable. To these, the formalism is the matter of interest, and they are happy to produce theorems and tables to their purpose.

While conceding that from time to time, both the nominalist and idealist approaches listed above have their uses, we shall in this book, be concerned largely with what would appear to be a middle ground: Namely, we shall usually be working with models, but with a view that the models themselves must be improved whenever it is feasible to do so. To us, simulation will provide a device for working with models, testing models, and building new models. So then, to us, simulation will be a kind of paradigm for realistic evolutionary modeling. At present, simulation is used by many as an adjuvant for dealing with old modeling techniques, say, the numerical approximation to pointwise evaluation of a differential equation. In the future, the author believes that simulation-based modeling will be at least as important as some of the older summarization models, such as differential equations. One will go directly from postulates and data to estimation and approximation without intervening classical summarization models. This would amount to something resembling a paradigm shift in the sense of Kuhn. It is a very big deal indeed to be able to say: "If our assumptions are correct, then here is a program for simulating a host of possible realizations with a variety of frequencies." At present, most simulations still consist of assists in dealing with older modeling summarizations. That is changing. To a large extent, the future of science will belong to those willing to make the shift to simulation-based modeling. This book has been written with such readers in mind.

In acknowledging support for this book, I start with thanks to my thesis advisor, John Tukey, who taught his students continually to question and modify preconceived notions of reality, and the late Elizabeth Tukey, whose graciousness and kindness will always be remembered by her husband's students. Then, I would like to acknowledge the support of the Army Research Office (Durham) under DAAH04-95-1-0665 and DAAD19-99-1-0150 for this work. Among my colleagues in Army science, I would particularly like to thank Robert Launer, Jagdish Chandra, Malcolm Taylor, Barry Bodt, and Eugene Dutoit. At the Polish Academy of Science, I would like to thank Jacek Koronacki. At Rice, I would like to thank Ed Williams, Katherine Ensor, Marek Kimmel, David Scott, Martin Lawera, Diane Brown, Tres Schwalb, Tony Elam, Sidney Burrus, Michael Carroll, Keith Baggerly, Dennis Cox, Patrick King, Peter Olofsson, Roxy Cramer, Mary Calizzi, and John Dobelman. At the University of Texas M.D. Anderson Cancer Center, I would like to thank Barry Brown and Neely Atkinson. At the University of South Carolina, I would like to thank Webster West. At Princeton, I would like to thank Stuart Hunter and the late Geoffrey Watson. At the Rand Corporation, I would like to thank Marc Elliott. I also wish to thank Steven Boswell of Lincoln Laboratories, and James Gen-

tle from Visual Numerics and George Mason University. At John Wiley &
Sons, I would like to thank my editor Stephen Quigley.

   Finally, and most importantly, I would like to thank my wife, Ewa Ma-
jewska Thompson, for her love and encouragement.

James R. Thompson

Houston, Texas
Easter, 1999

Simulation

Chapter 1

The Generation of Random Numbers

1.1 Introduction

There are many views as to what constitutes simulation. To the statistician, simulation generally involves randomness as a key component. Engineers, on the other hand, tend to consider simulation as a deterministic process. If, for example, an engineer wishes to simulate tertiary recovery from an oil field, he or she will probably program a finite element approximation to a system of partial differential equations. If a statistician attacked the same problem, he or she might use a random walk algorithm for approximating the pointwise solution to the system of differential equations.

In the broadest sense, we may regard the simulation of a process as the examination of any emulating process simpler than that under consideration. The examination will frequently involve a *mathematical model*, an oversimplified mathematical analogue of the real-world situation of interest. The related simpler process might be very close to the more complex process of interest. For example, we might simulate the success of a proposed chain of 50 grocery stores by actually building a single store and seeing how it progressed. At a far different level of abstraction, we might attempt to describe the functioning of the chain by writing down a series of equations to approximate the functioning of each store, together with other equations to approximate the local economies, and so on. It is this second level of abstraction that will be of more interest to us.

It is to be noted that the major component of simulation is neither stochasticity nor determinism, but rather, analogy. Needless to say, our visions of reality are always other than reality itself. When we see a forest, it is really a biochemical reaction in our minds that produces something to which we relate the notion of *forest*. When we talk of the *real world*, we really talk of perceptions of that world which are clearly other than that world but are (hopefully) in strong correlation with it. So, in a very real sense, analogy is "hardwired" into the human cognitive system. But to carry analogy beyond that which it is instinctive to do involves a learning process more associated with some cultures than with others. And it is the ability of the human intellect to construct analogies that makes modern science and technology a possibility. Interestingly, like so many other important advances in human thought, the flowering of reasoning by analogy started with Socrates, Plato, and Aristotle. Analogy is so much a part of Western thinking that we tend to take it for granted. In simulation we attempt to enhance our abilities to analogize to a level consistent with the tools at our disposal.

The modern digital computer, at least at the present time, is not particularly apt at analogue formulations. However, the rapid digital computing power of the computer has enormous power as a device complementary to the human ability to reason by analogy. Naturally, during most of the scientific epoch, the digital computer simply did not exist. Accordingly, it is not surprising that most of science is still oriented to methodologies in the formulation of which the computer did not play an intimate part.

The inventor of the digital computer, John von Neumann, created the device to perform something like random quadrature rather than to change fundamentally the precomputer methodology of modeling and analogy. And indeed, the utilization of the computer by von Neumann was oriented toward being a fast calculator with a large memory. This kind of mindset, which is a carryover of modeling techniques in the precomputer age, led to something rather different from what I call simulation, namely the *Monte Carlo method*.

According to this methodology, we essentially start to work on the abstraction of a process (through differential equations and the like) as though we had no computer. Then, when we find difficulties in obtaining a closed form solution, we use the computer as a means of facilitating pointwise function evaluation.

One conceptual difference between simulation and Monte Carlo in this book will be that simulation will be closer to the model of the system underlying the data. However, there is no clear demarcation between the Monte Carlo method on the one hand and simulation on the other. As we shall see later, a fuller utilization of the computer frequently enables us to dispense with abstraction strategies suitable to a precomputer age.

As an example of the fundamental change that the modern digital computer makes in the modeling process, let us consider a situation where we wish to examine particles emanating from a source in the interior of an irregular and heterogeneous medium. The particles interact with the medium by collisions with it and travel in an essentially random fashion.

The classical approach for a regular and symmetrically homogeneous medium would be to model the aggregate process by looking at differential equations that track the average behavior of the microaxioms governing the progress of the particles. For an irregular and nonsymmetrically homogeneous medium, the Monte Carlo investigator would attempt to use random walk simulations of these differential equations with pointwise change effects for the medium. In other words, the Monte Carlo approach would be to start with a precomputer age methodology and use the computer as a means for random walk implementations of that methodology.

If we wish to use the power of the digital computer more fully, we can go immediately from the microaxioms to random tracking of a large number of the particles. It is true that even with current computer technology, we will still not be in a position to deal with 10^{16} particles. However, a simulation dealing with 10^5 particles is both manageable and probably sufficient to make very reasonable conjectures about the aggregate of the 10^{16} particle system. In distinguishing between simulation and the Monte Carlo method, we will, in the former, be attempting the modeling in the light of the existence of a computer that may take us very close indeed to a precise emulation of the process under consideration. Clearly, then, *simulation* is a moving target. The faster the computer and the larger the storage, the closer we can come to a true simulation.

1.2 The Generation of Random Uniform Variates

Many simulations will involve some aspect of randomness. Theorem 1.1 shows that at least at the one-dimensional level, randomness can be dealt with if only we can find a random number generator from the uniform distribution on the unit interval $\mathcal{U}(0,1)$.

Theorem 1.1. Let X be a continuous random variable with distribution function $F(\cdot)$ [i.e., let $F(x) = P(X \leq x)$]. Consider the random variable $Y = F(x)$. Let the distribution function of Y be given by $G(y) = P(Y \leq y)$. Then Y is distributed as $\mathcal{U}(0,1)$.

Proof

$$G(y) = P(Y \leq y) = P(F(x) \leq y) = P(x \leq F^{-1}(y)) = y \qquad (1.1)$$

since $P(x \leq F^{-1}(y))$ is simply the probability that X is less than or equal to that value of X than which X is less y of the time. This is precisely the distribution function of the uniform distribution on the unit interval. This proves the theorem. •

For the simulator, Theorem 1.1 has importance rivaling that of the central limit theorem, for it says that all that is required to obtain a satisfactory random number generator for any continuous one-dimensional random variable, for which we have a means of inverting the distribution function, is a good $\mathcal{U}(0,1)$ generator. This is conceptually quite easy. For example, we might have an electrical oscillator in which a wavefront travels at essentially the speed of light in a linear medium calibrated from 0 to 1 in increments of, say, 10^{-10}. Then, we simply sense the generator at random times that an observer will pick. Aside from the obvious fact that such a procedure would be prohibitively costly, there seems to be a real problem with paying the observer and then reading the numbers into the computer. Of course, once it were done, we could use table look-up forever, being sure never to repeat a sequence of numbers once used.

Realizing the necessity for a generator that might be employed by the computer itself, without the necessity of human intervention except, perhaps, at the start of the generation process, von Neumann developed such a scheme. He dubbed the generator he developed the *midsquare method*. To carry out such a procedure, we take a number of, say, four digits, square it, and then take the middle four digits, which are used for the generator for the next step. If using base 10 numbering, we simply put a decimal before the first of the four digits. Let us show how this works with the simple example below.

We start with $X_0 = 3333$. Squaring this, we obtain 11<u>1088</u>89. Taking the middle four digits, we have $X_1 = 1088$. Squaring 1088 we have 1<u>1837</u>44. This gives $X_2 = 8374$, and so on. If we are using base 10, this gives us the string of supposed $\mathcal{U}(0,1)$ random variates.

The midsquare method is highly dependent on the starting value. Depending on the seed X_0, the generator may be terrible or satisfactory. Once we obtain a small value such as 0002, we will be stuck in a rut of small values until we climb out of the well. Moreover, as soon as we obtain 0, we have to obtain a new starter, since 0 is not changed by the midsquare operation.

Examinations of the midsquare method may be rather complicated mathematically if we are to determine, for example, the *cycle length*, the length of the string at which it starts to repeat itself. Some have opined that since this generator was used in rather crucial computations concerning nuclear reactions, civilization is fortunate that no catastrophe came about as a result of its use. As a matter of fact, for reasonable selections of *seeds* (i.e., starting values), the procedure can be quite satisfactory for most applications. It is, however, the specificity of behavior based on starting values that makes the method rather unpopular.

The midsquare method embodies more generally many of the attributes of *random number generators* on the digital computer. First, it is to be noted that such generators are not really random, since when we see part of the string, given the particular algorithm for a generator, we can produce the rest of the string. We might decide to introduce a kind of randomness by using the time on a computer clock as a seed value. However, it is fairly clear that we need to obtain generators realizing that the very nature of realistic generation of random numbers on the digital computer will produce problems. Attempting to wish these problems away by introducing factors that are random simply because we do not know what they are is probably a bad idea. Knuth [7] gives an example of an extremely complex generator of this sort that would appear to give very random-looking strings of random numbers, but in fact easily gets into the rut of reproducing the seed value forever.

The maxim of dealing with the devil we know dominates the practical creation of random number generators. We need to obtain algorithms which are conceptually simple so that we can readily discern their shortcomings. The most widely used of the current random number generation algorithms is the *congruential random number generator*. The apparent inventor of the congruential random number generation scheme is D.H. Lehmer, who introduced the algorithm in 1951 [10].

First, we note how incredibly simple the scheme is. Starting with a seed X_0, we build up the succession of "pseudorandom" numbers via the rule

$$X_{n+1} = aX_n + b(\text{mod } m). \tag{1.2}$$

One of the considerations given with such a scheme is the length of the string length of pseudorandom numbers before we have the first repeat. Clearly, by the very nature of the algorithm, with its one-step memory, once we have a repeat, the new sequence will repeat itself exactly. If our only concern is the length of the cycle before a repeat, a very easy fix is available. Choosing $a = b = 1$ and $X_0 = 0$, we have for any m,

$$
\begin{aligned}
X_1 &= 1 \\
X_2 &= 2 \\
X_3 &= 3 \\
&\cdots \quad \cdots \\
X_{m-1} &= m - 1.
\end{aligned} \tag{1.3}
$$

Seemingly, then, we have achieved something really spectacular, for we have a string that does not repeat itself until we get to an arbitrary length of m. Of course, the string bears little resemblance to a random string, since it marches straight up to m and then collapses back to 1. We have to come up with a generator such that any substring of any length appears to be random. Let us consider what happens when we choose $m = 90$, $a = 5$, and $b = 0$. Then, if we start with $X_0 = 7$, we have

$$
\begin{aligned}
X_1 &= 35 \\
X_2 &= 85 \\
X_3 &= 65
\end{aligned}
$$

$$\begin{aligned}
X_4 &= 55 \\
X_5 &= 5 \\
X_6 &= 25 \\
X_7 &= 35.
\end{aligned}$$
(1.4)

It could be argued that this string appears random, but clearly its cycle length of six is far too short for most purposes. Our task will be to find a long cycle length that also appears random. The rather simple proof of the following theorem is due to Morgan [10].

Theorem 1.2. Let $X_{n+1} = aX_n + b(\text{mod } m)$.
Let $m = 2^k$, $a = 4c + 1$, b be odd.
Then the string of pseudorandom numbers so generated has cycle length $m = 2^k$.

Proof

Let $Y_{n+1} = aY_n + b$.
Without loss of generality, we can take $X_0 = Y_0 = 0$.
Then

$$\begin{aligned}
Y_1 &= aY_0 + b = b \\
Y_2 &= ab + b = b(1 + a) \\
Y_3 &= aY_2 + b = ab(a + a) + b = b(1 + a + a^2) \\
&\quad \cdots \quad \cdots \\
Y_n &= b(1 + a + a^2 + a^3 + \ldots + a^{n-1}).
\end{aligned}$$
(1.5)

We observe that $X_i = Y_i - h_1 2^k$.
Now suppose that $X_i = X_j$ for $i > j$.
We wish to show that $i - j \geq 2^k$.
Now, $Y_i - Y_j = b(a^j + a^{j+1} + \ldots + a^{i-1})$.
If $X_i = X_j$, then
$ba^j W_{i-j} = ba^j(1 + a + a^2 + \ldots + a^{i-j-1}) = h_2 2^k$,
where $W_n = 1 + a + a^2 + \ldots + a^{n-1}$ for $n \geq 1$.
To prove the theorem, we must show that W_{i-j} cannot equal an integer multiple of 2^k if $i - j < 2^k$, that is,
$W_{i-j} \neq h_3 2^k$ for $i - j < 2^k$.
We shall suppose first of all that $\underline{i - j \text{ is odd.}}$
Then $i - j = 2t + 1$ for $t \geq 0$.
(This is the place we use the fact that $a = 4c + 1$.)
$W_{2t} = (1 - a^{2t})/(1 - a) = [(1+4c)^{2t} - 1]/[4c] = [(1+4c)^t - 1][(1+4c)^t + 1]/(4c)$
$= [(1 + 4c)^t + 1] \left[\sum_{i=1}^{t} (4c)^{i-1} \binom{t}{i} \right].$

But $1 + (1 + 4c)^t = 2 + 4c \sum_{i=1}^{t} (4c)^{i-1} \binom{t}{i}.$

So, $W_{2t+1} = W_{2t} + a^{2t}$ is odd, since a is odd.

Hence, if $i - j$ is odd, then $W_{i-j} \neq h_3 2^k$, since W_{i-j} is odd.

Next we wish to consider the case where $\underline{i - j \text{ is even.}}$

If $i - j$ is even, then there exists an s such that $i - j = \alpha 2^s$, for some odd integer α.

$$
\begin{aligned}
W_{i-j} &= W_{\alpha 2^s} \\
&= 1 + a + \ldots + a^{\alpha 2^{s-1}-1} + a^{\alpha 2^{s-1}} + \ldots + a^{\alpha 2^s - 1} \\
&= W_{\alpha 2^{s-1}} + a^{\alpha 2^{s-1}} \left(1 + a + a^2 + \ldots + a^{\alpha 2^{s-1}-1} \right) \\
&= W_{\alpha 2^{s-1}} + a^{\alpha 2^{s-1}} W_{\alpha 2^{s-1}} \\
&= W_{\alpha 2^{s-1}} \left(1 + a^{\alpha 2^{s-1}} \right).
\end{aligned}
\tag{1.6}
$$

Similarly, we have

$$
W_{\alpha 2^{s-1}} = W_{\alpha 2^{s-2}} \left(1 + a^{\alpha 2^{s-2}} \right).
\tag{1.7}
$$

Continuing the decomposition,

$$
\begin{aligned}
W_{i-j} &= W_{\alpha 2^{s-2}} \left(1 + a^{\alpha 2^{s-2}} \right) \left(1 + a^{\alpha 2^{s-1}} \right) \tag{1.8} \\
&= W_{\alpha} \left(1 + a^{\alpha} \right) \left(1 + a^{\alpha 2} \right) \ldots \left(1 + a^{\alpha 2^{s-1}} \right).
\end{aligned}
$$

Recalling that

$$
1 + (4c + 1)^j = 2 + 4c \sum_{i=1}^{j} \binom{j}{i} (4c)^i,
\tag{1.9}
$$

we see that $W_{i-j} = W_{\alpha} \gamma 2^s$. Note that we have shown in the first part of the proof that for α odd, W_{α} is also odd. Furthermore, we note that the product of terms such as $1 + even$ is also odd, so γ is odd. Thus if $W_{i-j} = h_2 2^k$, we must have $s = k$. •

The following more general theorem is stated without proof.

Theorem 1.3. Let $X_{n+1} = aX_n + b \bmod(m)$. Then the cycle of the generator is m if and only if
(i) b and m have no common factor other than 1.
(ii) $(a - 1)$ is a multiple of every prime number that divides m.
(iii) $(a - 1)$ is a multiple of 4 if m is a multiple of 4.

So far, we have seen how to construct a sequence of arbitrarily long cycle length. Obviously, if we wish our numbers to lie on the unit interval, we will simply divide the sequence members by m. Thus, the jth random number would be X_j/m. It would appear that there remains the design problem of selecting a and b to give the generator seemingly random results. To do this, we need to examine congruential generators in the light of appropriate perceptions of randomness. We address this issue in the next section.

1.3   Latticing and Other Problems

In a sense, it is bizarre that we ask whether a clearly deterministic sequence, such as one generated by a *congruential random number generator*, is random. Many investigators have expressed amazement at early "primitive" schemes, such as those of von Neumann, devised for random number generation. As a matter of fact, I am personally unaware of any tragedy or near tragedy caused by any of these primitive schemes (although I have experienced catastrophes myself when I inadvertently did something putatively wrong, such as using, repetitively, the same seed to start runs of, supposedly different, sequences of congruential random numbers). The fact is, of course, that the issue is always decided on the basis of what are the minimal requirements for a sequence of random numbers.

Let us note that having generators with long intervals before numbers are repeated is not sufficient. For example, suppose that we decided to use the generator

$$X_{n+1} = X_n + .000000001, \text{ where } X_0 = 0. \tag{1.10}$$

Such a generator will give us 10^9 numbers between 0 and 1 with never a repeat. Clearly, however, it is totally unsatisfactory, since it creeps slowly from 0 to 1 in steady increments. It is true that after one billion numbers are generated, we will have the number of points generated in an interval between 0 and 1 equal to one billion times the interval length, as we should. But, if only, say, 100,000 points are generated, there will be no points at all in the interval $[.0001,1]$. Probably, no one would use such a generator. A modest criterion would be that even for a small number of points generated, say N, we should observe that the total number of points in an interval of length ϵ should be roughly equal to ϵN. Practically speaking, all the congruential random number generators in use seem to have this property.

Suppose, however, that we are employing a congruential random number generator to give points in the unit hypercube of dimension greater than 1. That could mean, for example, that if we are generating points in two dimensions, we could use a congruential random number generator in such a fashion that the first number in a string would give us the first dimension of a double, the second would give the second dimension of the double. Then the third number in a string would give us the first dimension of a second double, the fourth, the second dimension, and so on.

For many applications, it will be sufficient if we can show that for any small volume, say ϵ, of a hypercube of unit volume, for a large number of generated random numbers, say N, the number of these falling in the volume will be approximately ϵN. But suppose it turned out that there were regions of the hypercube in which we never obtained any points, regardless of the number of points generated. Such behavior is observed for the once popular RANDU generator of IBM. A little work reveals the following relation [7]:

$$x_{i+1} = (6x_i - 9x_{i-1}) \bmod (2^{31}). \tag{1.11}$$

Such a relationship between successive triples is probably not disastrous for most applications, but it looks very bad when graphed from the proper view. Using D^2 Software's MacSpin, we observe two views of this generator, one, in Figure 1.1, seemingly random, the other in Figure 1.2, very much not.

$$x_{i+1} = (2^{16} + 3)x_i \bmod (2^{31}). \tag{1.12}$$

As a matter of fact, congruential random number generators generally have the problem of latticing. This holds even if we try to be clever by using a different generator for each dimension, or perhaps having one generator which randomly samples from each of M generators to pick a number for each dimension. One reasonable way to lessen this difficulty might be to use a generator that minimizes the maximum distance between two lattices. If this is achieved, then even though we will have vast empty regions (in fact, it is obvious that all the random numbers generated by a congruential random number generator must lie in a set of Lebesgue measure zero in the unit hypercube), it would be very difficult to conceive of a realistic situation where this might cause practical difficulty.

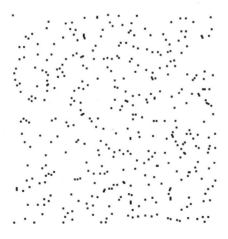

Figure 1.1. RANDU in Two Dimensions.

Figure 1.2. RANDU Spun to Show Latticing.

Accordingly, Knuth [7] suggests as a means of minimizing the difficulties associated with such latticing that one choose values of a, b, and m to give large values of

$$C_k = \frac{\pi^{k/2} \nu_k^k}{(k/2)! m}, \tag{1.13}$$

where the wavenumber ν_k is given as the solution to

$$\nu_k = \min \sqrt{s_1^2 + s_2^2 + \ldots + s_k^2} \tag{1.14}$$

subject to

$$s_1 + s_2 a + s_3 a^2 + \ldots + s_k a^{k-1} = 0 \mod(m). \tag{1.15}$$

Knuth suggests that we restrict ourselves to generators with $C_k \geq 1$ for $k = 2$, 3 and 4. One such random generator, where $a = 5^{15}$, $m = 2^{35}$, $b = 0$, has $C_2 = 2.02$, $C_3 = 4.12$, and $C_4 = 4.0$.

To motivate Knuth's argument, following Bratley, Fox and Schrage [3] we note that, for a family of parallel hyperplanes in k-space, we can write for any member

$$\alpha_1 X_1 + \alpha_2 X_2 + \ldots + \alpha_k X_k = mq \text{ for } q = 0, 1, 2, \ldots. \tag{1.16}$$

Now, for each family of equispaced parallel hyperplanes using the generator

$$X_{i+1} = a X_i \mod(m), \tag{1.17}$$

we observe that one hyperplane passes through the origin. So, to find the distance between adjacent hyperplanes, we can simply find the length of the line segment from the origin to the next hyperplane in the family. Thus, we can minimize the function

$$f(X_1, X_2, \ldots, X_k) = X_1^2 + X_2^2 + \ldots + X_k^2 \qquad (1.18)$$

subject to the constraint

$$\phi(X_1, X_2, \ldots, X_k) = \alpha_1 X_1 + \alpha_2 X_2 + \ldots + \alpha_k X_k = m. \qquad (1.19)$$

Using the Lagrange multiplier technique, this gives us

$$
\begin{aligned}
2X_1 + \lambda\alpha_1 &= 0 & (1.20) \\
2X_2 + \lambda\alpha_2 &= 0 & (1.21) \\
\ldots & \quad \ldots & (1.22) \\
2X_k + \lambda\alpha_k &= 0. & (1.23)
\end{aligned}
$$

Thus, the distance between adjacent hyperplanes is given by

$$\frac{m}{\sqrt{\alpha_1^2 + \alpha_2^2 + \ldots + \alpha_k^2}}. \qquad (1.24)$$

So, to minimize the distance between hyperplanes, we maximize

$$\sqrt{\alpha_1^2 + \alpha_2^2 + \ldots + \alpha_k^2}. \qquad (1.25)$$

Now, let us assume that one of the generated points, $X_i = 1$. Then, $X_{i+1} = a$, $X_{i+2} = a^2 \bmod(m)$, and so on. The point in k-space $(X_i, X_2, \ldots, X_k)$ must be on one of the hyperplanes. So

$$\alpha_1 X_1 + \alpha_2 X_2 + \ldots + \alpha_k X_k = \alpha_1 + a\alpha_2 + \ldots + a^{k-1}\alpha_k = mq \qquad (1.26)$$

for some integer q. Substituting for the α_i, s_i, we have the wavenumbers ν_k indicated. We note here that the variables to be optimized are a and m. We note the problem of trying to obtain the solution of some linear combination of the C_k, say, is nontrivial, and, generally speaking, the relative weights to be used in such a combination are likely to be arbitrary. Accordingly, Knuth's rule of thumb, namely obtaining reasonable separations for lattices in dimensions two through four, is likely to be satisfactory for most applications.

As a practical matter, the generator of Lewis, Goodman and Miller [8–9] has been subjected to 30 years of close scrutiny and seems generally satisfactory for most purposes. It is given, simply by

$$X_{n+1} = 7^5 X_n \bmod(2^{31} - 1). \qquad (1.27)$$

It has been the basic generator in the IMSL statistical package [5]. It is used as by Press et alia as a "minimal standard generator" in the *Numerical*

Recipes in Fortran [11]. There are three basic random number generators given in *Numerical Recipes in Fortran* (ran0, ran1, ran2) of increasing degrees of complexity, all based on the Lewis, Goodman and Miller algorithm. The third, ran2, is presented with a cash offer by Press et alia to anyone who can provide an example where it does not produce random-seeming numbers.

In a sense, the problem of latticing has been dealt with rather satisfactorily and is not one to which we should devote much further time. There is still another problem, however, which can cause difficulties in high-dimensional situations.

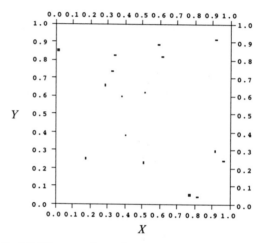

Figure 1.3. 16 Nonrandom-Seeming Random Numbers.

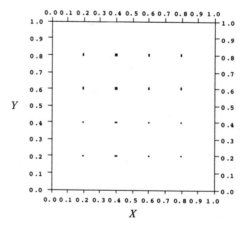

Figure 1.4. 16 Regular Mesh Points.

For small sample sizes, random numbers do not usually look very random. Using the random number generator from the SYSTAT package, we note, in Figure 1.3 that the 16 points generated do not look particularly random. Some areas of the unit square are devoid of points, whereas in others points seem to be rather close together.

Naturally, we could attempt to solve this problem by using a regular mesh as shown in Figure 1.4. The regular mesh approach achieves the goal of distributing points rather evenly over the unit square, but clearly the points so generated give no appearance of being random. Should this concern us? Certainly, if we are generating, say, random times at which orders are received at a mail order company and the random times required to fill them, the regular mesh approach would probably be very bad.

There are, of course, problems in which the regular mesh approach is used routinely. Suppose, for example, we wish to obtain a numerical approximation to

$$I = \int_0^1 \int_0^1 f(x,y)dxdy. \tag{1.28}$$

Then a fairly standard procedure would be to approximate I by

$$I \approx \frac{1}{(N+1)^2} \sum_{i=0}^{N} \sum_{j=0}^{N} f\left(\frac{i}{N}, \frac{j}{N}\right). \tag{1.29}$$

In lower-dimensional situations where we can perhaps identify the special anomalies that f might present, this would probably be satisfactory. Consider, however, the case where

$$f(x,y) = |\sin(2\pi Kx)\sin(2\pi Ky)|, \text{ with } K \text{ a multiple of } N. \tag{1.30}$$

Then the regular mesh approach would incorrectly estimate I by zero. The problem with using a regular mesh for such things as numerical quadrature is that the functions to be integrated are generally human-made and hence prone to regularities that may coincide with the regular mesh. If, as an alternative to the regular mesh, we used a random one of size $(N+1)^2$ using a congruential random number generator, such synchronicities of function regularity and mesh regularity are unlikely to occur. But as we have seen in Figure 1.3, there is still the problem that if N is too small, we might well neglect regions where the function is particularly large.

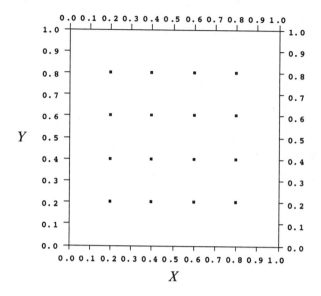

Figure 1.5. 16 Halton Random Numbers.

Many, of Cartesian mindset, tend to regard the regular mesh as having one of the properties of randomness. To these, random points are rather like particles each with the same positive charge, which when dropped into the hypercube, assume configurations based on the notion that each particle is repelled from the others. To solve the problem of creating a sequence of points in higher dimensions which look random without having the potentially destructive synchronicities of a regular mesh, Halton [4] has proposed the following recursive strategy:

$$p_r(r^n) = \frac{1}{r^{n+1}}, \text{ for } n = 0, 1, 2, \ldots;$$ (1.31)
$$p_r(r^n + j) = p_r(r^n) + p_r(j), \text{ for } j = 1, 2, \ldots, r^{n+1} - r^n - 1.$$

We consider in Figure 1.5 the use of $(p_2(i), p_3(i))$ for $i = 1, 2, \ldots, 16$. The greater regularity of the Halton sequence over that obtained from the congruential number generator is observed. The contrast between the congruential random number generator approach and that of the Halton sequence as we increase the number of generated doubles to 50 is shown in Figures 1.6 and 1.7, respectively.

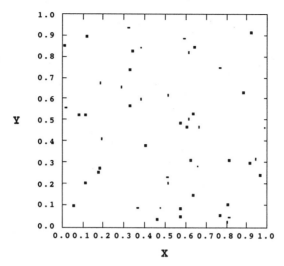

Figure 1.6. 50 Random Numbers.

In Table 1.1, we note that the $p_2(i)$ column has a property that might cause it difficulty for many applications: Namely, it moves upward and falls back in cycles of two. Similarly, the $p_3(i)$ column exhibits cyclicity of order three. Such properties might well cause spurious results if we were simulating orders into a warehouse. So it would appear that the use of Halton sequences in preference to congruential random numbers would buy one desirable property at the expense of losing another kind.

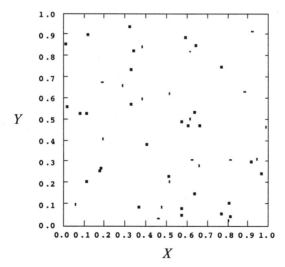

Figure 1.7. 50 Halton Numbers.

	Table 1.1			
	Congruential		Halton	
i	X	Y	$p_2(i)$	$p_3(i)$
1	0.5079	0.2295	0.5000	0.3333
2	0.1798	0.2523	0.2500	0.6667
3	0.3849	0.5952	0.7500	0.1111
4	0.7688	0.0541	0.1250	0.4444
5	0.5151	0.6200	0.6250	0.7777
6	0.9665	0.2434	0.3750	0.2222
7	0.9123	0.2948	0.8750	0.5555
8	0.2900	0.6579	0.0625	0.8888
9	0.8153	0.0425	0.5625	0.0370
10	0.3296	0.7372	0.3125	0.3704
11	0.5925	0.8859	0.8125	0.7037
12	0.0141	0.8483	0.1875	0.1481
13	0.3458	0.8226	0.6875	0.4815
14	0.6133	0.8153	0.4375	0.8148
15	0.9176	0.9095	0.9375	0.2593
16	0.4066	0.3778	0.0312	0.5926
17	0.8055	0.0188	0.5312	0.9259
18	0.4588	0.0296	0.2812	0.0741
19	0.1931	0.4053	0.7812	0.4074
20	0.6038	0.4681	0.1562	0.7407
21	0.8126	0.3097	0.6563	0.1852
22	0.6287	0.3059	0.4063	0.5185
23	0.3813	0.8386	0.9063	0.8519
24	0.5136	0.2009	0.0938	0.2963
25	0.0225	0.5592	0.5938	0.6296
26	0.6138	0.4995	0.3438	0.9630
27	0.1176	0.5262	0.8438	0.0123
28	0.5753	0.0803	0.2188	0.3457
29	0.0610	0.0997	0.7188	0.6790
30	0.1885	0.6729	0.4688	0.1235
31	0.9428	0.3149	0.9688	0.4568
32	0.1819	0.2662	0.0156	0.7901
33	0.1209	0.8979	0.5156	0.2346
34	0.5764	0.4865	0.2656	0.5679
35	0.3671	0.0845	0.7656	0.9012
36	0.6570	0.2787	0.1406	0.0494
37	0.8797	0.6278	0.6406	0.3827
38	0.1184	0.2001	0.3906	0.6790
39	0.6433	0.8453	0.8906	0.1605
40	0.3205	0.9359	0.0781	0.4938
41	0.5756	0.0492	0.5781	0.8272
42	0.6386	0.1475	0.3281	0.2716
43	0.7687	0.7440	0.8281	0.6049
44	0.8113	0.1037	0.2031	0.9383
45	0.3286	0.5669	0.7031	0.0864
46	0.4795	0.0869	0.4531	0.4198
47	0.6373	0.5300	0.9531	0.7531
48	0.9888	0.4633	0.0469	0.1975
49	0.0818	0.5248	0.5469	0.5309
50	0.6653	0.4675	0.2969	0.8642

There are ways that we can obtain the *positive charge* property of Halton sequences without obtaining the obvious cyclicity of the ones shown here. But then, one will be able to find some other nonrandom-seeming property of the generator and so on. Indeed, almost any system one might employ for generating random numbers on a digital computer is likely to be, in fact, deterministic. Hence, it will surely be found to exhibit some kind of nonrandom behavior if we look for it long enough. It is possible to try random number generation by, for example, reading off a high-frequency oscillator at regular low-frequency intervals. Is it worth the bother to come up with such hybridized analog–digital devices? Probably not, for most purposes.

Experience shows that we are probably well served to use one of the popular congruential number generators. The cataloging of the potential shortcomings of such generators is usually well documented. Hundreds of papers have been written on congruential random number generation, the associated difficulties, and ways to minimize these. By this time, for most purposes, the fairly standard procedures utilized in most software packages are quite satisfactory. For high-dimensional quadrature, Halton sequences provide a powerful alternative. However, interesting though the problem of congruential random number generation may be (and it represents just one of the rebirths of the venerable area of number theory), we shall not dwell much further on it.

Problems

1.1. One simple test to determine whether a random number generator is giving random values from a distribution with distribution function $F(x)$ is that of Kolmogorov and Smirnov [12]. Using the random number generator, obtain a sample of size n. Order the sample with x_r being the rth smallest observation. Define the *sample distribution function* S_n to be

$$
\begin{aligned}
S_n(x) &= 0 \text{ for } x < x_{(1)} \\
&= \frac{r}{n} \text{ for } x_r \le x_{r+1} < x_{r+1} \\
&= 1 \text{ for } x_n \le x.
\end{aligned}
$$

Then, for n larger than 80, it can be shown that if the x's are truly coming from F, then, with probability .99,

$$
\sup_x |S_n(x) - F(x)| < \frac{1.6276}{\sqrt{n}} .
$$

Use the random generator in (1.27) to generate a sample of size 1000 from the uniform distribution on the unit interval. Here,

$$
\begin{aligned}
F(x) &= 0 \text{ if } x < 0 \\
&= x \text{ if } 0 \le x \le 1 \\
&= 1 \text{ if } x > 1.
\end{aligned}
$$

Does your sample pass the Kolmogorov–Smirnov test?

1.2. We know from the central limit theorem that, under very general conditions, if we have a random sample, say $x_1, x_2, \ldots, x_n$ from a distribution with mean μ and standard deviation σ, then

$$
\bar{x} = \frac{1}{n} \sum_{j=1}^{n} x_j
$$

will, for n sufficiently large, approximately have the normal (Gaussian) distribution with mean μ and standard deviation $\sigma/\sqrt{n}$. Now, the density function of a normal variate with mean zero and standard deviation 1 is given by

$$f(x) = \frac{1}{\sqrt{2\pi}} \exp\left(-\frac{x^2}{2}\right).$$

The cumulative distribution function is not available in closed form. However, there are many excellent approximations, for example [1]

$$
\begin{aligned}
P(x) &= \int_{-\infty}^{x} \frac{1}{\sqrt{2\pi}} \exp\left(-\frac{t^2}{2}\right) dt. \\
&\approx 1 - \frac{1}{2}(1 + c_1 x + c_2 x^2 + c_3 x^3 + c_4 x^4)^{-4} + \epsilon(x),
\end{aligned}
$$

where

$$
\begin{aligned}
c_1 &= .196854 & c_3 &= .000344 \\
c_2 &= .115194 & c_4 &= .019527 \\
\epsilon(x) &< 2.5 \times 10^{-4}.
\end{aligned}
$$

Using the generator from Problem 1.1 (recalling that for the uniform distribution on $[0,1]$, $\mu = .500$ and $\sigma = 1\sqrt{12}$), use sample means of sizes 10, 20, and 50 to obtain a normal variable generator with mean zero and standard deviation 1. Examine how well such a generator passes the Kolmogorov test for a normal random variate.

1.3. A $\chi^2(n-1)$ random variable can be obtained as the sum of the squares of n independent normal variates with mean zero and unit standard deviation, that is,

$$\chi^2 = z_1^2 + z_2^2 + \ldots + z_n^2$$

has cumulative distribution function given by

$$F(\chi^2) = \int_0^{\chi^2} \frac{1}{\Gamma((n-1)/2)2^{(n-1)/2}} w^{(n-1)/2} \exp(-w/2)dw.$$

Using the central limit theorem based generator in Problem 1.3, generate 1000 realizations of a χ^2 variate with 8 degrees of freedom.

1.4. One of the oldest statistical tests is Karl Pearson's *goodness of fit*. According to this test, we divide the X-axis into *bins*. According to theory, out of n observations, we expect the ith bin to contain np_i observations. So, if the underlying density function is given by $g(x)$, we would then expect the probability of a random observation falling between b_i and b_{i+1} to be

$$p_i = \int_{b_i}^{b_{i+1}} g(x)dx.$$

Then if X_i is the number of observations falling between b_i and b_{i+1}, if the intervals are disjoint, for n large,

$$Q_{k-1} = \sum_{i=1}^{k} \frac{(X_i - np_i)^2}{np_i}$$

has approximately the χ^2 distribution with $k - 1$ degrees of freedom. Use this test in Problems 1.1–1.3 to see whether the generators proposed pass the goodness-of-fit test at the 1% significance level.

1.5. Some procedures for generating random numbers may be inefficient from the standpoint of computing time but are conceptually quite easy for the user to implement. And, of course, we should be in the business of making things easy for people, even at the expense of possible computational inefficiencies. Among the most generated random numbers are those from the Gaussian distribution. Let us suppose that we have two uniform independent random numbers U_1 and U_2 from $\mathcal{U}[0, 1]$. Define X_1 and X_2 by

$$
\begin{aligned}
X_1 &= \sqrt{(-2\ln(U_1)}\cos(2\pi U_2) & (1.32)\\
X_2 &= \sqrt{(-2\ln(U_1)}\sin(2\pi U_2) \ .
\end{aligned}
$$

Prove that X_1 and X_2 are independently distributed as $\mathcal{N}(0, 1)$. This procedure (first proposed by Box and Muller [2]) carries computational baggage in that the computer takes quite a lot of flops to approximate transcendental functions such as logarithms, sines and cosines. However, the present power of desktop computers minimizes the importance of such considerations.

1.6. A t variate with $n - 1$ degrees of freedom can be generated from a random sample of size n from a normal distribution with mean μ and variance σ^2. We then compute the sample mean $\overline{X}$ and sample variance s^2. Then

$$t = \frac{\overline{X} - \mu}{s/\sqrt{n}}$$

is a t variate with $n-1$ degrees of freedom. Using the generator in Problem 1.5, create histograms of t variates using 1000 simulations of sizes $n= 2, 10$ and 60.

1.7. The generation of random samples from discrete distributions where the variable can take the integer values $0, 1, 2, \ldots$ can be carried out by the following algorithm:

1. Compute the probabilities $p_1, p_2, \ldots$.

2. Generate a random uniform variate u from $\mathcal{U}[0, 1]$.

3. If $u \leq p_0$, assign the variable to event 0, otherwise,

4. If $p_0 < u \leq p_0 + p_1$, assign the variable to event 1; otherwise,

5. If $p_0 + p_1 < u \leq p_0 + p_1 + p_2$, assign the variable to event 2, otherwise, and so on.

(a) Generate a sample of size 1000 from a binomial distribution with $p = .2$ and $n = 20$.

(b) Simulate a deal of bridge hands to four players.

1.8. In the generation of random numbers, some tasks are conceptually easy but can require a fair amount of computation time. One of these is the acceptance−rejection method of von Neumann [13]. For example, if we wish to sample uniformly from a 10-dimensional sphere of unit radius, we can simply use (1.27) 10 times to sample from the uniform distribution on the unit interval. Then we can multiply each number by 2 and subtract 1 from it to obtain samples from the interval [-1, +1]. The resulting 10-tuple,

$$(2u_1 - 1, 2u_2 - 1, \ldots, 2u_{10} - 1) = (U_1, U_2, \ldots, U_{10}),$$

is a candidate for a point inside the 10-sphere. The only requirement is that

$$U_1^2 + U_2^2 + \ldots + U_{10}^2 < 1.$$

See how many times you must use the one-dimensional generator in order to obtain a sample of size 100 from the 10-sphere. *Hint*: The volume of an n-dimensional sphere of radius r is

$$V_n = \frac{(\pi)^{n/2}}{\Gamma(n/2 + 1)} r^n.$$

1.9. Generate p independent random variables, $X_1, X_2, \ldots, X_p$ from $\mathcal{N}(0, 1)$. Let

$$\begin{aligned}
Z_1 &= a_{11}X_1 \\
Z_2 &= a_{21}X_1 + a_{22}X_2 \\
&\ldots \qquad \ldots \\
Z_p &= a_{p1}X_1 + a_{p2}X_2 + \ldots + a_{pp}X_p.
\end{aligned} \qquad (1.33)$$

Prove that $\mathbf{Z}$ is a p-dimensional Gaussian random variable with zero means and with covariance Σ given by

$$\Sigma = \begin{pmatrix}
\sigma_{11} & \sigma_{12} & \cdots & \sigma_{1p} \\
\sigma_{12} & \sigma_{22} & \cdots & \sigma_{2p} \\
\cdots & \cdots & \cdots & \cdots \\
\sigma_{1p} & \sigma_{2p} & \cdots & \sigma_{pp}
\end{pmatrix}, \qquad (1.34)$$

where

$$
\begin{aligned}
\sigma_{11} &= a_{11}^2 \\
\sigma_{12} &= a_{11}a_{21} \\
\cdots \quad &\quad \cdots \\
\sigma_{1p} &= a_{11}a_{p1} \\
\sigma_{22} &= a_{21}^2 + a_{22}^2 \\
\cdots \quad &\quad \cdots \\
\sigma_{pp} &= a_{p1}^2 + a_{p2}^2 + \ldots + a_{pp}^2.
\end{aligned}
\tag{1.35}
$$

Finally, show that

$$
\begin{aligned}
W_1 &= Z_1 + \mu_1 \\
W_2 &= Z_2 + \mu_2 \\
\cdots \quad &\quad \cdots \\
W_p &= Z_p + \mu_p
\end{aligned}
\tag{1.36}
$$

is a p-dimensional normal random variable with mean

$$
\mu = \begin{pmatrix} \mu_1 \\ \mu_2 \\ \cdots \\ \mu_p \end{pmatrix}
$$

and covariance matrix

$$
\Sigma = \begin{pmatrix}
\sigma_{11} & \sigma_{12} & \cdots & \sigma_{1p} \\
\sigma_{12} & \sigma_{22} & \cdots & \sigma_{2p} \\
\cdots & \cdots & \cdots & \cdots \\
\sigma_{1p} & \sigma_{2p} & \cdots & \sigma_{pp}
\end{pmatrix}.
$$

References

[1] Abramowitz, M. and Stegun, L.A. (1968). *Handbook of Mathematical Functions*. Washington,D.C.: National Bureau of Standards, 932.

[2] Box, G.E.P. and Muller, M.A. (1958). "A note on the generation of random normal deviates," *Ann. Math. Stat.*, **29**, 610.

[3] Bratley, P., Fox, B.L., and Schrage, L.E. (1987). *A Guide to Simulation*. New York: Springer Verlag, 209–210.

[4] Halton, J.H. (1960). "On the efficiency of certain quasi-random sequences of points in evaluating multi-dimensional integrals," *Numer. Math.*, **2**, 163–171.

[5] *IMSL STAT/LIBRARY.* Houston, Texas: Visual Numerics.

[6] Kennedy, W.J. and Gentle, J.E. (1980). *Statistical Computing.* New York: Marcel Dekker.

[7] Knuth, D.E. (1981). *The Art of Computing,* **2**, *Seminumerical Algorithms.* Reading, Mass.: Addison-Wesley, 84–100.

[8] Lewis, P.A.W., Goodman, A.S., and Miller, J.M. (1969). "A pseudorandom number generator for the System/360," *IBM Syst. J.,* 136–146.

[9] Lewis, P.A.W. and Orav, E.J. (1989). *Simulation Methodology for Statisticians, Operations Analysts, and Engineers.* Pacific Grove, Calif.: Wadsworth. 65–95.

[10] Morgan, B.J.T. (1984). *Elements of Simulation.* London: Chapman & Hall, 65–74.

[11] Press, W.H., Teukolsky, S.A., and Vetterling, W.T., and Flannery, B.P. (1992). *Numerical Recipes in FORTRAN.* New York: Cambridge University Press, 266–290.

[12] Stuart, A. and Ord, J.K. (1991). *Kendall's Advanced Theory of Statistics,* **2**. New York: Oxford University Press, 1187–1188.

[13] von Neumann, J. (1951). "Various techniques used in connection with random digits," *U.S. NBS Applied Mathematics Series* **12**, 36–38.

Chapter 2

Random Quadrature

2.1 Introduction

We have noted the fact that the principal aspect of simulation is analogy-based model formulation rather than stochasticity. We further noted that simulation is much more useful when used prospectively in the model formulation rather than as some fix of a formulation prepared as though no simulation would ever take place. As an example of simulation as an afterthought we consider now the use of simulation in the evaluation of definite integrals in one dimension. In other words, we wish to use an appropriate randomization technique for evaluating integrals of the form

$$S = \int_a^b g(x)\,dx, \tag{2.1}$$

where $g(x)$ is a continuous function. Very frequently, such an integration would be handled by an appropriate quadrature procedure such as Simpson's rule with n intervals:

$$\int_a^b g(x)dx \approx \frac{h}{3}(g(0)+g(b)+4[g(h)+g(3h)+...]+2[g(2h)+g(4h)+...]), \tag{2.2}$$

where $h = (b-a)/n$.

There are many reasons why a deterministic quadrature might be unsatisfactory. For example, the function might be quite large over a portion of the interval relative to that in the rest of the interval. Particularly when we go to higher dimensions, the issue of grid selection becomes a problem. Note that if we use the four-dimensional version of Simpson's quadrature, then a mesh of 100 on a side would require function evaluation at 100,000,000 points, a lengthy task even on a very fast computer. Suppose that we wished, in the four-dimensional case discussed above, to go to a coarse mesh, say one with five points on a side. This could assuredly be done, since only 625 function evaluations would be required. However, we should ask ourselves how we would choose the five points on a side. Most of us would let two of these be a_j and b_j. This would surely place a great deal of emphasis on what went on near the periphery of the four-dimensional parallelepiped. Or, we could decide, systematically, to exclude the vertices, thus missing whatever interesting might be going on near the periphery. The problem is that a human mind probably abstracted the function g, and a human mind will also pick the deterministic mesh. Quite frequently, these selections will be such as to miss systematically a great deal if we insist on using deterministic quadrature. One of the more intuitive schemes of random quadrature is that of *hit-or-miss Monte Carlo*.

2.2 Hit-or-Miss Monte Carlo

Let us evaluate the integral in (2.1), being guided by Figure 2.1. We shall sample from the two-dimensional density:

$$
\begin{aligned}
u(x,y) &= \frac{1}{c(b-a)} \text{ if } a < x < b, 0 < y < c \\
&= 0, \text{ otherwise.}
\end{aligned}
$$

We shall sample from u. If a point lies below $g(x)$, we shall increase the counter of successes, x, by one and the counter of tries by one. If the point lies above the curve, however, we shall increase the counter of tries by one. Thus, our estimate for equation (2.1) is given after n tries by

$$
\hat{\theta}_1 = \frac{x}{n}. \tag{2.3}
$$

Clearly, X is a binomial random variable with expectation given by

$$
p = \frac{S}{c(b-a)} \tag{2.4}
$$

and variance

$$
\sigma^2 = \frac{p(1-p)}{n}. \tag{2.5}
$$

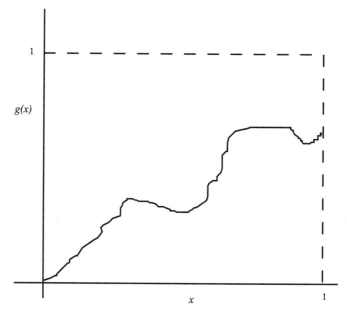

Figure 2.1. Hit or Miss Monte Carlo.

This enables us to arrive at the approximate 95% confidence interval

$$\int_a^b g(x)\,dx \approx \frac{x}{n} \pm 2\sqrt{\frac{x/n(1 - x/n)}{n}}. \tag{2.6}$$

In the remainder of our discussion, we shall make the simplifying assumption that $a = 0$ and $b = 1$. Further, we shall see to it that $g(x)$ is scaled such that in the unit interval, $0 \le g(x) \le 1$. To gain a better feeling for the hit-or-miss method, let us consider a short simulation for the case where $g(x) = \exp(-x^2)$ as shown in Table 2.1.

Table 2.1

j	x_j	y_j	$e^{-x_j^2}$
1	.279	.914	.925
2	.024	.506	1.000
3	.603	.311	.696
4	.091	.302	.999
5	.774	.183	.550

We note that since the sample value of y is less each time than the corresponding value of e^{-x^2}, we arrive at the estimate $\hat{\theta}_1 = 1$, and a statement to the effect that the 95% confidence interval is 1 ± 0, not a very reassuring result, considering that the actual value of θ here is .7569. The Monte Carlo approach generally gives us a ready means of obtaining confidence sets for our estimates, but particularly for small sample sizes, these are to be taken with a grain of salt.

The example above gives us a feeling as to the generally poor results one can expect from it hit-or-miss Monte Carlo. The answer must always be some integer multiple of $1/N$. Hence, only for large N should we expect any sort of accuracy under the best of circumstances.

We have effectively taken a problem dealing with randomness in one dimension and artificially added randomness in another dimension. Looking at the above example formally, we have been dealing with

$$\hat{\theta}_1 = \int_0^1 \int_0^1 h(x, y) \, dy \, dx \qquad (2.7)$$

where

$$
\begin{aligned}
h(x, y) &= 1 \text{ if } y < e^{-x^2} \\
&= 0 \text{ if } y \geq e^{-x^2}
\end{aligned}
$$

and

$$\int_0^1 h(x, y) \, dy = e^{-x^2}.$$

Adding additional randomness in this way is to be avoided in most situations. We note that in the example above we could have been even more unwise by going to a random process in the unit cube, and so on.

2.3   Sample Mean Monte Carlo

Returning to our basic problem, namely the estimation of

$$\theta = \int_0^1 g(x) \, dx, \qquad (2.8)$$

we can simply take a sample of size N from a uniform distribution on the unit interval to obtain the sample average estimator

$$\hat{\theta}_2 = \frac{1}{N} \sum_{j=1}^N g(x_j). \qquad (2.9)$$

Clearly,

$$E(\hat{\theta}_2) = \frac{1}{N} \sum_{j=1}^N E(g(x_j)) = \frac{1}{N} \theta = \theta. \qquad (2.10)$$

The variance of $\hat{\theta}_2$ is, of course,

$$\text{Var}(\hat{\theta}_2) = \frac{\text{Var}(g(x_j))}{N} = \frac{\int_0^1 (g(x) - \theta)^2 \, dx}{N}. \qquad (2.11)$$

Naturally, we shall not have the exact expression for $\text{Var}(\hat{\theta}_2)$ at our disposal. However, we shall be able to employ the sample variance:

$$s_{\hat{\theta}_2}^2 = \frac{1}{N} \frac{\sum_{j=1}^{N}(g(x_j) - \hat{\theta}_2)^2}{N-1}. \tag{2.12}$$

Once again, let us return to our case study of estimating

$$\theta = \int_0^1 e^{-x^2} dx$$

using the evaluation of the integrand at five points selected at random from the uniform distribution on the unit interval as shown in Table 2.2.

		Table 2.2	
j	x_j	$e^{-x_j^2}$	$(e^{-x_j^2})^2$
1	.279	.925	.819
2	.024	1.000	1.000
3	.603	.696	.485
4	.091	.999	.998
5	.774	.550	.302
	Sums	4.170	3.604

This gives us

$$\hat{\theta}_2 = \frac{4.170}{5} = .803 \tag{2.13}$$

$$s_{\hat{\theta}_2}^2 = \frac{3.604 - 3.225}{5(4)} = .019. \tag{2.14}$$

This gives us as the 95% confidence interval for the true value of θ,

$$\theta = \hat{\theta}_2 \pm 2s_{\hat{\theta}_2} = .803 \pm .276. \tag{2.15}$$

Let us compare, in general, the variance associated with hit-or-miss Monte Carlo and that associated with sample mean Monte Carlo for a given value of N.

$$\begin{aligned}
\text{Var}(\hat{\theta}_1) - \text{Var}(\hat{\theta}_2) &= \frac{\theta(1-\theta)}{N} - \frac{1}{N}\int_0^1 (g(x) - \theta)^2 dx \\
&= \frac{\theta}{N} - \frac{\theta^2}{N} - \frac{1}{N}\int_0^1 g^2(x)dx + \frac{\theta^2}{N} \tag{2.16} \\
&= \frac{\theta}{N} - \frac{1}{N}\int_0^1 g^2(x)dx \\
&= \frac{1}{N}\int_0^1 [g(x) - g^2(x)]dx \\
&\geq 0, \text{ since we have assumed } 0 \leq g(x) \leq 1.
\end{aligned}$$

Thus, we see that sample mean Monte Carlo is generally preferred to hit-or-miss Monte Carlo.

Before going to more sophisticated means of carrying out *stochastic quadrature*, it is appropriate to recall the medieval tale about the magic soup nail.

> One day, a poor widow preparing her simple dinner of barley soup was disturbed by a knock on the door of her hut. Opening the door, she saw a peddler. The peddler said, "Good woman, I seek naught from thee save a pot of boiling water. I will take this magic nail and immerse it in the water. Then, I trow, we will have the most delicious soup in the world. I will take but two bowls of the soup, and give thee the rest."
>
> This sounded fair enough, so the poor widow allowed the peddler the use of the pot of water already bubbling on the stove. With great formality, the peddler produced a rusty nail and immersed it twice in the pot. Smacking his lips in anticipation, he said, "Ah, what a treat we shall have." Then, after a moment, a frown crossed his brow. "Wonderful though this soup is, it would be the better still had we a bit of barley to thicken it." The widow produced her cup of barley which the peddler added to the pot. He then took out his magic nail and dipped it into the pot again. "Good woman," he said, "did you have but a piece of pork, we would be blessed the more." The widow produced a piece of salted ham from her meager store. In short order, the peddler conned the widow into producing leeks, salt, and herbs. And the soup, the widow decided, was as good as she had tasted. Following the meal, the peddler generously sold the widow the nail for a silver shilling.

The point of the story in the current context is that improvements over sample mean Monte Carlo may require a fair amount of additives to the basic barley soup of SMMC. These additives may sometimes be readily at hand, but generally they will cost us one of our most important resources—people time. The speed of modern desktop computers is such that in most cases, sample mean Monte Carlo will be the method of choice. It is intuitive, quick to code, and provides easy estimation of error levels.

2.4   Control Variate Sampling

The problem of evaluating θ arises due to the fact that we have no ready knowledge of the value of the integral. Suppose, however, there is a function $h(x)$ such that $h(x) \approx g(x)$ for $0 \le x \le 1$ and that we can integrate $h(x)$ over the unit interval. Then, we can use Monte Carlo on the second of the two integrals

$$\int_0^1 h(x)dx + \int_0^1 [g(x) - h(x)]dx. \qquad (2.17)$$

Due to the fact that g and h are approximately equal on the interval, the value of the second integral, and hence of the sum of the two integrals (the first being known and thus with zero variance) is less than the integral of g.

We shall use as the approximating function to e^{-x^2}, the quadratic $1-.7x^2$. First of all, we note that

$$\int_0^1 (1 - .7x^2)dx = 1 - \frac{.7}{3} = .766. \qquad (2.18)$$

Taking our example of five sample points on the unit interval, we have Table 2.3.

Table 2.3

j	x_j	$e^{-x_j^2}$	$1 - .7x_j^2$	$g(x_j) - h(x_j)$
1	.279	.925	.945	-.020
2	.024	1.000	.996	.004
3	.603	.696	.745	-.049
4	.091	.999	.994	.005
5	.774	.550	.580	-.030
			Sum	-.090

This gives us, then, as our estimate for θ, simply

$$\hat{\theta}_3 = .766 - \frac{.09}{5} = .748. \qquad (2.19)$$

In general, estimates obtained by this method will be of the form

$$\hat{\theta}_3 = \int_0^1 h(x)dx + \frac{1}{N}\sum_{j=1}^N [g(x_j) - h(x_j)]. \qquad (2.20)$$

Clearly,

$$\begin{aligned} E(\hat{\theta}_3) &= \int_0^1 h(x)dx + \frac{1}{N}\sum_{j=1}^N [E(g(x_j)) - E(h(x_j))] \\ &= \int_0^1 h(x)dx + \frac{1}{N}[N\int_0^1 g(x)dx - N\int_0^1 h(x)dx] \qquad (2.21) \\ &= \int_0^1 g(x)dx = \theta. \end{aligned}$$

Next

$$
\begin{aligned}
\mathrm{Var}(\hat{\theta}_3) &= E[(\hat{\theta}_3 - \theta)^2] \\
&= \frac{1}{N}[\mathrm{Var}(g(x) - h(x))] \\
&= \frac{1}{N}[\mathrm{Var}(g(x)) + \mathrm{Var}(h(x)) - 2\mathrm{Cov}(g(x), h(x))] \\
&= \mathrm{Var}(\hat{\theta}_2) + \frac{1}{N}[\mathrm{Var}(h(x)) - 2\mathrm{Cov}(g(x), h(x))]. \quad (2.22)
\end{aligned}
$$

Thus, if $2\mathrm{Cov}(g(x), h(x)) > \mathrm{Var}(h(x))$, then $\mathrm{Var}(\hat{\theta}_3) < \mathrm{Var}(\hat{\theta}_2)$. In such a case, the control variate estimator bests the sample mean estimator.

2.5   Importance Sampling

Returning again to the estimation of θ, let us consider multiplication and division under the integral sign by $k(x)$ where k is a probability density function on the unit interval, that is, $k(x) > 0$ for $0 \le x \le 1$ and $\int_0^1 k(x)dx = 1$. We then represent θ as follows:

$$
\theta = \int_0^1 \frac{g(x)}{k(x)}k(x)dx = E[\frac{g(y)}{k(y)}] = \int_0^1 [\frac{g(y)}{k(y)}]dK(y). \quad (2.23)
$$

Sampling from the distribution function K, we can obtain N independent y_j's from $K(y)$ and obtain the Monte Carlo estimate

$$
\hat{\theta}_4 = \frac{1}{N}\sum_{j=1}^{N}\frac{g(y_j)}{k(y_j)}. \quad (2.24)
$$

Since

$$
E\left[\frac{g(y)}{k(y)}\right] = \int_0^1 \frac{g(y)}{k(y)}dK(y) = \int_0^1 \frac{g(x)}{k(x)}k(x)dx = \theta, \quad (2.25)
$$

$$
E(\hat{\theta}_4) = \theta. \quad (2.26)
$$

The variance of $\hat{\theta}_4$ is given by

$$
\begin{aligned}
\mathrm{Var}(\hat{\theta}_4) &= \frac{1}{N}\mathrm{Var}\left[\frac{g(x)}{k(y)}\right] \\
&= \frac{1}{N}\left(E\left[\frac{g(y)}{k(y)}\right]^2 - \theta^2\right) \\
&= \frac{1}{N}\left[\int_0^1 \frac{g^2(y)}{k^2(y)}dK(y) - \theta^2\right] \\
&= \frac{1}{N}\left[\int_0^1 \frac{g^2(y)}{k(y)}dy - \theta^2\right]. \quad (2.27)
\end{aligned}
$$

The best that can be done is to have

$$k(x) = \frac{g(x)}{\int_0^1 g(x)dx} = \frac{g(x)}{\theta}, \tag{2.28}$$

for in this case

$$
\begin{aligned}
\mathrm{Var}(\hat{\theta}_4) &= \frac{1}{N}\left[\int_0^1 \frac{g^2(x)dx}{g(x)/\theta} - \theta^2\right] \\
&= \frac{1}{N}\left(\left[\int_0^1 g(x)dx\right]^2 - \theta^2\right) \\
&= 0. \tag{2.29}
\end{aligned}
$$

The selection of $k(x)$ should be such that it is $g(x)$ normalized to be a density function. Of course, we do not know $\int g(x)dx$. Returning to our example, since we have seen that $h(x) = 1 - .7x^2$ is fairly close to e^{-x^2} on the unit interval, the normalized version of $h(x)$ should work reasonably well. We have seen that $\int_0^1 (1 - .7x^2)dx = .766$. This leads us to sample from the density function for $0 \le y \le 1$.

$$k(y) = \frac{1}{.766}(y - .233y^3). \tag{2.30}$$

Thus, we have as the cumulative distribution fundtion, (cdf)

$$
\begin{aligned}
K(x) &= \quad 0 \text{ if } x \le 0 \tag{2.31} \\
&= \frac{1}{.767}(x - .233x^3) \text{ if } 0 < x \le 1 \\
&= \quad 1 \text{ if } x > 1.
\end{aligned}
$$

Now, $w = K(y)$ has the uniform distribution on the unit interval. Thus

$$P[K(y) \le w] = w = P[x \le y] = \frac{1}{.767}(y - .233y^3). \tag{2.32}$$

Hence, we may sample u from $U(0,1)$ and solve the equation

$$\frac{.233y^3}{.767} - \frac{1}{.767}y + u = 0. \tag{2.33}$$

Since we do not actually know the optimal $k(x)$, we need not carry the iteration out to a high degree of approximation.

Returning to our sample of five sample uniform deviates on the unit interval, we have in Table 2.4,

Table 2.4					
j	u_j	y_j	$g(y_j)$	$h(y_j)$	$g(y_j)/h(y_j)$
1	.279	.210	.961	1.27	.758
2	.024	.021	1.000	1.31	.789
3	.603	.451	.600	1.12	.535
4	.091	.069	.999	1.30	.770
5	.774	.541	.766	1.04	.735
				Sum	3.587

This leaves us with the estimate

$$\hat{\theta}_4 = \frac{3.587}{5} = .717. \tag{2.34}$$

2.6   Stratification

In the evaluation of θ, we may find it convenient to write the integral as

$$\int_0^1 g(x)dx = \int_{a_1}^{b_1} g(x)dx + \int_{a_2}^{b_2} g(x)dx + ... + \int_{a_m}^{b_m} g(x)dx, \tag{2.35}$$

where $0 = a_1 < b_1 = a_2 < b_2 = a_3 < ... < a_m < b_m = 1$. By using sample mean Monte−Carlo on each of the m pieces, we take samples of size N_l from each interval where $N_1 + N_2 + ... + N_l + ... + N_m = N$.

Let us call

$$\theta_l = \int_{a_l}^{b_l} g(x)dx = (b_l - a_l) \int_{a_l}^{b_l} g(x)\frac{dx}{b_l - a_l}. \tag{2.36}$$

To obtain a random variable from $U(a_l, b_l)$, we sample y from $U(0,1)$ and let $x = a_l + (b_l - a_l)y$. In this way, we obtain estimates of the form

$$\hat{\theta}_l = \frac{b_l - a_l}{N_l} \sum_{j=1}^{N_l} g(x_{l_j}). \tag{2.37}$$

Our estimate of θ will be

$$\hat{\theta}_5 = \sum_{l=1}^m \hat{\theta}_l = \sum_{l=1}^m (b_l - a_l) \sum_{j=1}^{N_l} g(x_{l_j}). \tag{2.38}$$

Now,

$$\begin{aligned}
\text{Var}(\hat{\theta}_5) &= \sum_{l=1}^m \text{Var}(\hat{\theta}_l) = \sum_{l=1}^m \frac{(b_l - a_l)^2}{N_l{}^2} \sum_{j=1}^{N_l} \text{Var}(g(x_{l_j})) \\
&= \sum_{l=1}^m \frac{(b_l - a_l)^2}{N_l} \text{Var}(g(x_l)). \tag{2.39}
\end{aligned}$$

To achieve optimal allocation of samples from stratum to stratum, we need to minimize $\text{Var}(\hat{\theta}_5)$ subject to the constraint

$$\sum_{l=1}^{m} N_l = N. \qquad (2.40)$$

Treating the N_l's as continuous variables, we have

$$\frac{\partial(\text{Var}(\hat{\theta}_5))}{\partial N_l} = -\frac{(b_l - a_l)^2}{N_l{}^2}\text{Var}[g(x_l)] \qquad (2.41)$$

subject to the constraint

$$\frac{\partial N}{\partial N_l} = 1. \qquad (2.42)$$

Using the method of Lagrange multipliers, we have

$$-(b_1 - a_1)^2 N_1{}^{-2}\text{Var}[g(x_1)] + \lambda \;\; = 0$$
$$-(b_2 - a_2)^2 N_2{}^{-2}\text{Var}[g(x_2)] + \lambda \;\; = 0$$
$$\cdots \quad \cdots$$
$$-(b_m - a_m)^2 N_m{}^{-2}\text{Var}[g(x_m)] + \lambda \;\; = 0.$$

This has the solution

$$\frac{N_l}{N_j} = \frac{b_l - a_l}{b_j - a_j} \frac{\sqrt{\text{Var}[g(x_l)]}}{\sqrt{\text{Var}[g(x_j)]}}. \qquad (2.43)$$

Thus, we achieve the optimal allocation of points per stratum by placing them in proportion to the standard deviation of g in each interval times the length of each interval. Naturally, knowing the variability of g precisely is likely to be as difficult as knowing the integral itself. However, we can use the rough rule of thumb that the points can generally be selected in rough proportion to the difference in the values of g at the endpoints of each interval.

We consider below the same example as for the other candidates for the estimation of θ. We shall divide the unit interval at intervals of width .2. Here, the function is not changing dramatically more in one interval than in any other, so we shall simply put one point in each interval as shown in Table 2.5.

Table 2.5				
j	u_j	x_j	$e^{-x_j^2}$	$\hat{\theta}_l$
1	.279	.055	1.000	.200
2	.024	.205	.961	.192
3	.603	.521	.763	.153
4	.091	.618	.682	.136
5	.774	.955	.403	.081
			$\hat{\theta}_5$	.762

2.7   Antithetic Variates

As we are willing to assume more and more about the function whose definite integral we are attempting to evaluate, we can do better and better in the estimation of the integral. One dramatic example of such an improvement is that of antithetic variates due to Hammersley and Handscomb [2]. To introduce the procedure, let us suppose we have two unbiased estimators $\hat{\theta}$ and $\tilde{\theta}$. Then their average is also an unbiased estimator for θ, and

$$\text{Var}(\frac{\hat{\theta}+\tilde{\theta}}{2}) = \frac{1}{4}\text{Var}(\hat{\theta}) + \frac{1}{4}\text{Var}(\tilde{\theta}) + \frac{1}{2}\text{Cov}(\hat{\theta},\tilde{\theta}). \qquad (2.44)$$

Such an estimator will have smaller variance than either $\hat{\theta}$ or $\tilde{\theta}$ provided that

$$\frac{1}{4}\text{Var}(\hat{\theta}) + \frac{1}{4}\text{Var}(\tilde{\theta}) + \frac{1}{2}\text{Cov}(\hat{\theta},\tilde{\theta}) < \text{Min}(\text{Var}(\hat{\theta}), \text{Var}(\tilde{\theta})). \qquad (2.45)$$

Let us suppose without loss of generality that $\text{Var}(\hat{\theta}) < \text{Var}(\tilde{\theta})$. Then we require that

$$\text{Cov}(\hat{\theta},\tilde{\theta}) < \frac{3}{2}\text{Var}(\hat{\theta}) - \frac{1}{2}\text{Var}(\tilde{\theta}). \qquad (2.46)$$

If it should be the case that $g(x)$ is monotone, then one way to develop an estimator consisting of two negatively correlated parts is simply to use, on the basis of a sample of size N from the uniform distribution on the unit interval, the average of $\hat{\theta}$ and $\tilde{\theta}$, where

$$\hat{\theta} = \frac{1}{N}\sum_{j=1}^{N} g(x_j) \qquad (2.47)$$

and

$$\tilde{\theta} = \frac{1}{N}\sum_{j=1}^{N} g(1 - x_j). \qquad (2.48)$$

We note that if x is uniformly distributed on the unit interval, then so also is $y = 1 - x$. Turning to our standard example where $g(x) = e^{-x^2}$, we obtain Table 2.6.

Table 2.6				
j	x_j	$1 - x_j$	$g(x_j)$	$g(1 - x_j)$
1	.279	.721	.925	.595
2	.024	.971	1.000	.389
3	.603	.397	.696	.857
4	.091	.909	.999	.374
5	.774	.226	.550	.951
			$\hat{\theta} = .834$	$\tilde{\theta} = .633$

This gives us a pooled estimator of

$$\hat{\theta}_6 = \frac{1}{2}(\hat{\theta} + \tilde{\theta}) = .734. \tag{2.49}$$

Really substantial gains in efficiency are available using the antithetic variable technique. For example, consider

$$\hat{\theta}_6 = \frac{1}{N}\sum_{j=1}^{N}(\hat{\theta}_{1j} + \hat{\theta}_{2j} + \ldots + \hat{\theta}_{kj}), \tag{2.50}$$

where

$$\hat{\theta}_{lj} = (b_l - a_l)g(a_l + (b_l - a_l)x_j). \tag{2.51}$$

Then

$$E(\hat{\theta}_6) = \int_0^1 g(x)dx \tag{2.52}$$

and

$$\text{Var}(\hat{\theta}_6) = \frac{1}{N^2}[\sum_{j=1}^{N}\sum_{i=1}^{k}\text{Var}(\hat{\theta}_{ij}) + 2\sum_{j=1}^{N}\sum_{i=1}^{k-1}\sum_{l=i+1}^{j}\text{Cov}(\hat{\theta}_{ij}, \hat{\theta}_{lj})]. \tag{2.53}$$

If we let $b_l - a_l = 1/k$, then

$$\hat{\theta}_6 = \frac{1}{kN}\sum_{j=1}^{N}\sum_{i=1}^{k}g\left(\frac{x_i + (j-1)}{k}\right). \tag{2.54}$$

Then it can be shown that

$$\text{Var}(\hat{\theta}_6) = \frac{1}{N}\left(\frac{\Delta_0{}^2}{12k^2} + \frac{\Delta_1{}^2 - 2\Delta_0\Delta_2}{720k^4} + \frac{\Delta_2{}^2 - 2\Delta_1\Delta_3 + 2\Delta_0\Delta_4}{30240m^6}\right), \tag{2.55}$$

where

$$\Delta_j = \frac{d^j g}{dx^j}\Big|_{x=1} - \frac{d^j g}{dx^j}\Big|_{x=0} . \tag{2.56}$$

Clearly, it is desirable to make as many of the coefficients of the $1/k^{2j}$ terms vanish as is feasible. Let us first work on making Δ_0^2 vanish. Consider the transformation

$$\mathcal{T}_\alpha g(\xi) = \alpha g(\alpha\xi) + (1-\alpha)g[1 - (1-\alpha)\xi], \tag{2.57}$$

where ξ is $\mathcal{U}(0,1)$ and $0 < \alpha < 1$. Then

$$
\begin{aligned}
E[\mathcal{T}_\alpha g(\xi)] &= \alpha \int_0^1 g(\alpha\xi)d\xi + (1-\alpha)\int_0^1 g[1-(1-\alpha)\xi]d\xi \\
&= \alpha \int_0^\alpha g(0)\frac{dz}{\alpha} - (1-\alpha)\int_1^\alpha g(w)\frac{dw}{1-\alpha} \\
&= \int_0^\alpha g(z)dz + \int_\alpha^1 g(w)dw \\
&= \int_0^1 g(z)dz = \theta. \tag{2.58}
\end{aligned}
$$

Thus, it is clear that $\mathcal{T}_\alpha g(\xi)$ is an unbiased estimator for $\int_0^1 g(z)dz = \theta$. Let us consider the estimator

$$\mathcal{T}_{\alpha,j} g(\xi) = \frac{1}{k}\left[\alpha g\left(\frac{j}{k} + \frac{1}{k}\xi\alpha\right) + (1-\alpha)g\left(\frac{j+1}{k} - (1-\alpha)\frac{\xi}{k}\right)\right]. \tag{2.59}$$

Then

$$
\begin{aligned}
E[\mathcal{T}_{\alpha,j} g(\xi)] &= \frac{\alpha}{k}\int_0^1 g\left(\frac{j}{k} + \frac{\alpha}{k}\xi\right)d\xi + \frac{1-\alpha}{k}\int_0^1 g(\frac{j+1}{k} - (1-\alpha)\frac{\xi}{k})d\xi \\
&= \int_{\frac{j}{k}}^{\frac{j}{k}+\frac{\alpha}{k}} g(z)dz + \int_{\frac{j+\alpha}{k}}^{\frac{j+1}{k}} g(z)dz \\
&= \int_{\frac{j}{k}}^{\frac{j+1}{k}} g(z)dz. \tag{2.60}
\end{aligned}
$$

Let us consider, then, the estimator

$$\hat{\theta}_6 = \frac{1}{N}\sum_{i=1}^N \sum_{j=1}^k \mathcal{T}_{\alpha,j-1} g(x_i), \tag{2.61}$$

where x is $\mathcal{U}(0,1)$. From the preceding, it is clear that

$$E(\hat{\theta}_6) = \int_0^1 g(x)dx = \theta. \tag{2.62}$$

Returning to (2.56), we now seek a means whereby we may make

$$\Delta_0 = T[g(0)] - T_\alpha[g(1)] = 0. \tag{2.63}$$

Now,

$$\Delta_0 = T_\alpha g(\xi) = \alpha g(\alpha\xi) + (1 - \alpha)g[1 - (1 - \alpha)\xi]. \tag{2.64}$$

We require that

$$\alpha g(0) + (1 - \alpha)g(1) - \alpha g(\alpha) - (1 - \alpha)g(\alpha) = 0 \tag{2.65}$$

or

$$\alpha g(0) + (1 - \alpha)g(1) - g(\alpha) = 0$$

and

$$g(\alpha) = (1 - \alpha)g(1) + \alpha g(0).$$

Let us now apply the foregoing technique to the case where $g = e^{-x^2}$, $N = 1$, and $k = 4$. Let

$$e^{-\alpha^2} = (1 - \alpha)e^{-1} + \alpha$$

$$f(\alpha) = e^{-\alpha^2} + \alpha(e^{-1} - 1) - e^{-1} = 0$$

$$f'(\alpha) = -2\alpha e^{-\alpha^2} + e^{-1} - 1.$$

We use Newton's method to gain an approximate value of α to solve $f(\alpha) = 0$:

$$\alpha_{n+1} = \alpha_n - \frac{f(\alpha_n)}{f'(\alpha_n)}.$$

Starting with a first guess $\alpha_1 = .5000$, we have

$$\alpha_1 = .5000$$
$$\alpha_2 = .5000 - \frac{.0948}{-1.4109} = .5672$$
$$\alpha_3 = .5672 - \frac{-.0015}{-1.4544} = .5662$$
$$\alpha_4 = .5662 - \frac{-.0001}{-1.3679} = .5661.$$

Thus, we shall use

$$T_{\alpha,j}e^{-\xi^2} = \frac{1}{4}.5661\exp\left(-\left[\frac{j}{4} + \frac{.5661\xi}{4}\right]^2\right)$$
$$+ \frac{1}{4}.4339\exp\left(-\left[\frac{j+1}{4} - \frac{.4339\xi}{4}\right]^2\right). \tag{2.66}$$

Let us use the generated $\mathcal{U}(0,1)$ observation $\xi = .279$. Then

$$\hat{\theta}_6 = \frac{1}{4} \sum_{j=1}^{4} \mathcal{T}_{\alpha, j-1} e^{-\xi^2} = .7505. \tag{2.67}$$

The computation required here, even though only one random variable was generated, is greater than that required for sample means Monte Carlo with a sample of size 8. Nevertheless, we are extremely close to the correct value of θ, namely, .7469. We have, in (2.56) reduced the coefficient of the $1/k^2$ term to zero. It can be shown that we annihilate the coefficients of the $1/k^2$ and $1/k^4$ terms by replacing $\mathcal{T}_\alpha g(\xi)$ in (2.62) by

$$\frac{2}{3} \left[\mathcal{T}_\alpha g\left(\frac{\xi}{2}\right) + \mathcal{T}_\alpha g\left(\frac{1+\xi}{2}\right) \right] - \frac{1}{3} \left[\mathcal{T}_\alpha g(\xi) \right]. \tag{2.68}$$

The $1/k^2$, $1/k^4$ and $1/k^6$ terms coefficients may be annihilated by using

$$\frac{8}{21} \left[\mathcal{T}_\alpha g\left(\frac{\xi}{4}\right) + \mathcal{T}_\alpha g\left(\frac{1+\xi}{4}\right) + \mathcal{T}_\alpha g\left(\frac{2+\xi}{4}\right) + \mathcal{T}_\alpha g\left(\frac{3+\xi}{4}\right) \right]$$
$$- \frac{2}{7} \left[\mathcal{T}_\alpha g\left(\frac{\xi}{2}\right) + \mathcal{T}_\alpha g\left(\frac{1+\xi}{2}\right) \right] + \frac{1}{21} \mathcal{T}_\alpha g(\xi). \tag{2.69}$$

Although somewhat tedious, these transformations, may be well worth the trouble in many situations.

2.8   Least Squares Estimators

Let us consider the regression model

$$y_j = \beta_1 x_{1j} + \beta_2 x_{2j} + \ldots + \beta_k x_{kj} + \epsilon_j, \tag{2.70}$$

where the ϵ_j are independent and identically distributed with

$$\begin{aligned} E(\epsilon_j) &= 0 \\ \text{Var}(\epsilon_j) &= \sigma^2 \end{aligned}$$

and the $\{x_{i,j}\}$ are known real quantities. To estimate the regression coefficients $\{\beta_i\}$, we may minimize

$$S(\beta) = \sum_{j=1}^{n} (y_j - \beta_1 X_{1j} - \ldots - \beta_k x_{kj})^2. \tag{2.71}$$

In other words, minimize

$$S = \sum_{j=1}^{n} (y_{\text{observed},j} - y_{\text{predicted},j})^2. \tag{2.72}$$

This can be expressed in matrix notation by minimizing

$$S(\widehat{\boldsymbol{\beta}}) = (\boldsymbol{y} - \boldsymbol{X}\widehat{\boldsymbol{\beta}})'(\boldsymbol{y} - \boldsymbol{X}\widehat{\boldsymbol{\beta}}) \tag{2.73}$$

where

$$\boldsymbol{y} = \begin{bmatrix} y_1 \\ y_2 \\ \vdots \\ y_n \end{bmatrix} \; ; \; \widehat{\boldsymbol{\beta}} = \begin{bmatrix} \hat{\beta}_1 \\ \hat{\beta}_2 \\ \vdots \\ \hat{\beta}_n \end{bmatrix} \; ; \; \boldsymbol{\epsilon} = \begin{bmatrix} \hat{\epsilon}_1 \\ \hat{\epsilon}_2 \\ \vdots \\ \hat{\epsilon}_n \end{bmatrix}$$

$$E[\,\boldsymbol{\epsilon}] = \begin{bmatrix} 0 \\ 0 \\ \vdots \\ 0 \end{bmatrix} = \boldsymbol{0}$$

$$\mathrm{Var}[\boldsymbol{\epsilon}] = E[\boldsymbol{\epsilon}\boldsymbol{\epsilon}'] = \sigma^2 \mathbf{I}$$

$$\mathbf{X} = \begin{bmatrix} x_{11} & x_{12} & \cdots & x_{1n} \\ x_{21} & x_{22} & \cdots & x_{2n} \\ \vdots & \vdots & & \vdots \\ x_{n1} & x_{n2} & \cdots & x_{nn} \end{bmatrix} .$$

Imposing

$$\frac{\partial S}{\partial \hat{\beta}_1} = \frac{\partial S}{\partial \hat{\beta}_2} = \ldots = \frac{\partial S}{\partial \hat{\beta}_k} = 0, \tag{2.74}$$

we have

$$2\mathbf{X}'(\mathbf{y} - \mathbf{X}\,\widehat{\boldsymbol{\beta}}) = \boldsymbol{0} \tag{2.75}$$
$$\mathbf{X}'\mathbf{y} = (\mathbf{X}'\mathbf{X})\,\widehat{\boldsymbol{\beta}}.$$

Assuming that $X'X$ is nonsingular, we have

$$\widehat{\boldsymbol{\beta}} = (\mathbf{X}'\mathbf{X})^{-1}\mathbf{X}'\mathbf{y}. \tag{2.76}$$

Next, let us consider estimating several linear functions of the β_is:

$$\begin{aligned}
t_1 &= c_{11}\beta_1 + c_{21}\beta_2 + \ldots + c_{k1}\beta_k \tag{2.77} \\
t_2 &= c_{12}\beta_1 + c_{22}\beta_2 + \ldots + c_{k2}\beta_k \\
\ldots &= \ldots \\
t_m &= c_{1m}\beta_1 + c_{2m}\beta_2 + \ldots + c_{km}\beta_k
\end{aligned}$$

or

$$\mathbf{t} = \mathbf{c}\,\boldsymbol{\beta}. \tag{2.78}$$

We shall restrict our estimates to those that are linear in $\{y_j\}$, that is,

$$\widehat{\mathbf{t}} = \widehat{\mathbf{T}}\mathbf{y} \tag{2.79}$$

where

$$\hat{\mathbf{T}} = \begin{bmatrix} \hat{t}_{11} & \hat{t}_{21} & \cdots & \hat{t}_{n1} \\ \hat{t}_{12} & \hat{t}_{22} & \cdots & \hat{t}_{n2} \\ \vdots & \vdots & & \vdots \\ \hat{t}_{1m} & \hat{t}_{2m} & \cdots & \hat{t}_{nm} \end{bmatrix}.$$

Also, we shall require that $\hat{\mathbf{t}}$ be unbiased: that

$$E(\hat{\mathbf{t}}) = \mathbf{c}\boldsymbol{\beta}. \tag{2.80}$$

We now use the fact that

$$\mathbf{y} = \mathbf{X}\boldsymbol{\beta} + \boldsymbol{\epsilon}. \tag{2.81}$$

Then

$$\mathbf{t} = \mathbf{c}\boldsymbol{\beta} = E[\hat{\mathbf{T}}\mathbf{y}] = E[\hat{\mathbf{T}}(\mathbf{X}\boldsymbol{\beta} + \boldsymbol{\epsilon})] = \hat{\mathbf{T}}\mathbf{X}\boldsymbol{\beta}. \tag{2.82}$$

Thus

$$\hat{\mathbf{T}}\mathbf{X}\boldsymbol{\beta} = \mathbf{c}\boldsymbol{\beta} \tag{2.83}$$

or

$$\hat{\mathbf{T}}\mathbf{X} = \mathbf{c}.$$

The variance matrix of $\hat{\mathbf{t}}$ is given by

$$\mathbf{Var}(\hat{\mathbf{t}}) = E[(\hat{\mathbf{t}} - \mathbf{c}\boldsymbol{\beta})(\hat{\mathbf{t}} - \mathbf{c}\boldsymbol{\beta})']. \tag{2.84}$$

Now

$$\begin{aligned} \hat{\mathbf{t}} = \hat{\mathbf{T}}\mathbf{y} &= \hat{\mathbf{T}}(\mathbf{X}\boldsymbol{\beta} + \boldsymbol{\epsilon}) \\ &= \mathbf{c}\boldsymbol{\beta} + \hat{\mathbf{T}}\boldsymbol{\epsilon}. \end{aligned} \tag{2.85}$$

Hence

$$\hat{\mathbf{t}} - \mathbf{c}\boldsymbol{\beta} = \hat{\mathbf{T}}\boldsymbol{\epsilon}.$$

So

$$\mathbf{Var}(\hat{\mathbf{t}}) = E[\hat{\mathbf{T}}\boldsymbol{\epsilon}\boldsymbol{\epsilon}'\mathbf{T}'] = \sigma^2\hat{\mathbf{T}}\hat{\mathbf{T}}'. \tag{2.86}$$

We seek to minimize the diagonal elements of $\hat{\mathbf{T}}\hat{\mathbf{T}}'$. Now

$$\begin{aligned} \hat{\mathbf{T}}\hat{\mathbf{T}}' &= (\mathbf{c}(\mathbf{X}'\mathbf{X})^{-1}\mathbf{X}')(\mathbf{c}(\mathbf{X}'\mathbf{X})^{-1}\mathbf{X}')' \\ &\quad +(\hat{\mathbf{T}} - \mathbf{c}(\mathbf{X}'\mathbf{X})^{-1}\mathbf{X}')(\hat{\mathbf{T}} - \mathbf{c}(\mathbf{X}'\mathbf{X})^{-1}\mathbf{X}')'. \end{aligned} \tag{2.87}$$

$\hat{\mathbf{T}}$ appears only in the second term. The diagonal elements of both terms are nonnegative. Hence, the best we can do is to make the second term vanish. To achieve this we set

$$\hat{\mathbf{T}} = \mathbf{c}(\mathbf{X}'\mathbf{X})^{-1}\mathbf{X}', \tag{2.88}$$

giving

$$\hat{\mathbf{t}} = \mathbf{c}(\mathbf{X}'\mathbf{X})^{-1}\mathbf{X}'\mathbf{y} = \mathbf{c}\widehat{\boldsymbol{\beta}}, \tag{2.89}$$

where $\widehat{\beta}$ is the least squares estimate of β. The implications are clear. Namely, if we wish to estimate $c\widehat{\beta}$ in minimum variance unbiased linear fashion, we should simply use $c\widehat{\beta}$, whatever c is.

Since $\mathbf{Var}(\widehat{t}) = \sigma^2 \widehat{T}\widehat{T}'$, we have

$$\mathbf{Var}(\widehat{t}) = \sigma^2 \mathbf{c}(\mathbf{X}'\mathbf{X})^{-1}\mathbf{c}'. \tag{2.90}$$

In the case where the covariance matrix of ϵ is given by $\sigma^2\mathbf{V}$ (where nonsingular $\mathbf{V}$ is not necessarily the identity $\mathbf{I}$), a modification of the argument above gives for the minimum variance unbiased linear estimator for $\mathbf{c}\beta$:

$$\widehat{\mathbf{t}} = \mathbf{c}(\mathbf{X}'\mathbf{V}^{-1}\mathbf{X})^{-1}\mathbf{X}'\mathbf{V}^{-1}\mathbf{y}. \tag{2.91}$$

Moreover,

$$\mathbf{Var}(\mathbf{t}) = \sigma^2 \mathbf{c}(X'V^{-1}X)^{-1}c'. \tag{2.92}$$

In particular, if $\mathbf{c} = \mathbf{I}$,

$$\widehat{\mathbf{t}} = (\mathbf{X}'\mathbf{V}^{-1}\mathbf{X})^{-1}\mathbf{X}'\mathbf{V}^{-1}\mathbf{y}. \tag{2.93}$$

Now, let us consider the problem of estimating $\boldsymbol{\theta} = (\theta_1, \theta_2, \ldots, \theta_k)$, where

$$\theta_i = \int_0^1 g_i(x)dx.$$

Let us suppose that we have estimates $\{\widehat{t}_i\}_{l=1}^m$ such that

$$
\begin{aligned}
E(\widehat{t}_1) &= d_{11}\theta_1 + d_{12}\theta_2 + \ldots + d_{1k}\theta_k \\
E(\widehat{t}_2) &= d_{21}\theta_1 + d_{22}\theta_2 + \ldots + d_{2k}\theta_k \\
\ldots &= \ldots \\
E(\widehat{t}_m) &= d_{m1}\theta_1 + d_{m2}\theta_2 + \ldots + d_{mk}\theta_k.
\end{aligned}
\tag{2.94}
$$

We suppose we know the covariance matrix

$$
\mathbf{V} == \mathbf{V}(\widehat{\mathbf{t}}) =
\begin{bmatrix}
\mathrm{Var}(\widehat{t}_1, \widehat{t}_1) & \mathrm{Var}(\widehat{t}_1, \widehat{t}_2) & \ldots & \mathrm{Var}(\widehat{t}_1, \widehat{t}_m) \\
\mathrm{Var}(\widehat{t}_1, \widehat{t}_2) & \mathrm{Var}(\widehat{t}_2, \widehat{t}_2) & \ldots & \mathrm{Var}(\widehat{t}_2, \widehat{t}_m) \\
\vdots & \vdots & & \vdots \\
\mathrm{Var}(\widehat{t}_1, \widehat{t}_m) & \mathrm{Var}(\widehat{t}_2, \widehat{t}_m) & \ldots & \mathrm{Var}(\widehat{t}_m, \widehat{t}_m)
\end{bmatrix}.
\tag{2.95}
$$

We wish to obtain the minimum variance unbiased linear estimator for $(\theta_1, \theta_2, \ldots, \theta_k)$, that is, we wish to have

$$\widehat{\theta}_i = c_{i1}\widehat{t}_1 + c_{i2}\widehat{t}_2 + \ldots + c_{im}\widehat{t}_m, \tag{2.96}$$

where

$$\theta_i = E(\widehat{\theta}_i) = c_{i1}\sum_{l=1}^k d_{1l}\theta_l + \ldots + c_{im}\sum_{l=1}^k d_{ml}\theta_l \tag{2.97}$$

and $E(\hat{\theta}_i - \theta_i)^2$ is minimized for $l = 1, 2, \ldots, k$. That is, we shall select $\mathbf{c}$ to minimize

$$S(\mathbf{c}) = \sum_{i=1}^{k} E(\hat{\theta}_i - \theta_i)^2 \qquad (2.98)$$

subject to (2.97). Using Lagrange multipliers, we obtain

$$\hat{\boldsymbol{\theta}} = (\mathbf{D}'\mathbf{V}^{-1}\mathbf{D})^{-1}\mathbf{D}'\mathbf{V}^{-1}\hat{\mathbf{t}}. \qquad (2.99)$$

Even if $\mathbf{V}$ is not the true covariance matrix of $\{\hat{t}_l\}_{l=1}^{m}$, but merely an estimate, say $\hat{\mathbf{V}}$, we have

$$E(\hat{\boldsymbol{\theta}}) = \boldsymbol{\theta}. \qquad (2.100)$$

Once again, let us seek to estimate

$$\theta = \int_0^1 e^{-x^2} dx = \int_0^1 g(x) dx.$$

Let

$$\hat{t}_1 = \frac{1}{2}[g(\xi/2) + g(.5 + \xi/2)]$$

$$\hat{t}_2 = \frac{1}{2}[g(\xi) + g(1 - \xi)],$$

where ξ is $\mathcal{U}(01,1)$, as shown in Table 2.7.

Table 2.7

j	ξ_j	$g(.5\xi_j)$	$g(.5 + .5\xi_j)$	$g(\xi_j)$	$g(1 - \xi_j)$	$\hat{t}_{1j}$	$\hat{t}_{2j}$	$\hat{t}_{1j}\hat{t}_{2j}$
1	.279	.9807	.6643	.9251	.5946	.8225	.7598	.6249
2	.024	.9999	.7694	.9994	.3857	.8844	.6926	.6125
3	.603	.9131	.5260	.6952	.8542	.7196	.7747	.5575
4	.091	.9979	.7426	.9918	.4377	.8702	.7148	.6680
5	.774	.8609	.4553	.5493	.9502	.6581	.7498	.4934
					Sums	3.9548	3.6917	2.9104

This gives us

$$\hat{t}_1 = .7910 \qquad \hat{s}_{\hat{t}_1}^2 = \frac{1}{4}(3.1668 - 3.1281) = .00963$$

$$\hat{t}_2 = .7383 \qquad \hat{s}_{\hat{t}_2}^2 = \frac{1}{4}(2.7303 - 2.7259) = .00114$$

$$\mathrm{Cov}(\hat{t}_1, \hat{t}_2) = \frac{2.9104 - 2.9201}{4} = .00243$$

$$\hat{\mathbf{V}} = \begin{bmatrix} .00963 & -.00243 \\ -.00243 & .00140 \end{bmatrix}$$

$$\hat{\mathbf{V}}^{-1} = \frac{1}{|\hat{\mathbf{V}}|} \begin{bmatrix} .00114 & .00243 \\ .00243 & .00963 \end{bmatrix}$$

$$\mathbf{D} = \begin{bmatrix} 1 \\ 1 \end{bmatrix}$$

and

$$\hat{\theta} = (\mathbf{D}'\mathbf{V}^{-1}\mathbf{D})^{-1}\mathbf{D}'\mathbf{V}^{-1}\hat{\mathbf{t}} = .7503.$$

An approximation for the covariance matrix of $\hat{\theta}$ is given by

$$(\mathbf{D}'\hat{\mathbf{V}}^{-1}\mathbf{D})^{-1}/N.$$

In the above example we have

$$\mathrm{Var}(\hat{\theta}) \approx 6.4917 \times 10^{-5}$$

$$\sqrt{\mathrm{Var}(\hat{\theta})} \approx .0086.$$

So we should expect that

$$.7331 \le .7503 - 2(.0086) \le \theta \le .7503 + 2(.0086) = .7675.$$

2.9   Evaluation of Multidimensional Integrals

For most one-dimensional integrals, one can beat Monte Carlo procedures by some other device such as deterministic quadrature. For high-dimensional integrals, the picture changes. Consider, for example, the evaluation of

$$\int_0^1 \int_0^1 \int_0^1 \int_0^1 \int_0^1 \int_0^1 \frac{\exp[\sqrt{x_1} + x_2 x_3 + x_4^2 x_5]}{\sqrt{x_1 x_4 + x_3 x_6 + .01}} \Pi_{j=1}^6 dx_j. \qquad (2.101)$$

Careful study of the integrand would undoubtedly reveal the possibility of some clever quadrature procedure. However, without such an involved examination, we would be forced to use some naive quadrature procedure, requiring the imposition of a grid on the six-dimensional hypercube. If such a grid is imposed at intervals of width $1/10$ in each dimension, we must face 10^6 evaluations of the integrand.

However, Monte Carlo procedures can be used to obtain ballpark estimates quite easily. Consider, in general, the evaluation of integrals over the unit cube

$$\theta = \int_0^1 g(\mathbf{x})d\mathbf{x}. \qquad (2.102)$$

A simpleminded Monte Carlo strategy would sample $\mathbf{x}$ from the k-dimensional uniform distribution $\mathcal{U}(\mathbf{0}, \mathbf{1})$. This is easily accomplished by sampling each component from $\mathcal{U}(0, 1)$. Then, we have as an estimate for θ,

$$\hat{\theta} = \frac{1}{N} \sum_{j=1}^{N} g(\mathbf{x}_j). \qquad (2.103)$$

Defining

$$\text{Var}(g(\mathbf{x})) = \int_0^1 [g(\mathbf{x}) - \theta]^2 d\mathbf{x}, \tag{2.104}$$

then

$$\text{Var}(\hat{\theta}) = \frac{\text{Var}(g(\mathbf{x}))}{N}. \tag{2.105}$$

In any practical situation, we will not know $\text{Var}(g(\mathbf{x}))$. However, we may compute

$$s_g^2 = \frac{1}{N-1} \sum_{j=1}^N [g(\mathbf{x}_j) - \hat{\theta}]^2. \tag{2.106}$$

Assuming N is large enough for the central limit theorem to apply, we may say with 95 % confidence that

$$\hat{\theta} - 2\frac{s}{\sqrt{N}} < \hat{\theta} + 2\frac{s}{\sqrt{N}}. \tag{2.107}$$

Although

$$\text{Var}(\hat{\theta}) = \frac{1}{N} \text{Var}[g(\mathbf{x})]$$

whatever k may be, we must not infer that by the use of Monte Carlo we can escape unscathed from the curse of dimensionality.

Example. Let $g(\mathbf{x}) = x_1 + x_2 + \ldots + x_k$. Then

$$
\begin{aligned}
\text{Var}[g(\mathbf{x})] &= \int_0^1 \cdots \int_0^1 [x_1 - .5 + x_2 - .5 + \ldots + x_k - .5]^2 d\mathbf{x} \\
&= k \int_0^1 (x - .5)^2 dx = k\text{Var}(x_1).
\end{aligned}
$$

In general, we will be well advised to reduce the dimensionality of the problem whenever possible, either by analytically integrating out one or more of the dimensions or by exploiting some property of symmetry.

Example. In physics, integration over the unit 3-sphere is quite common. In particular, integrals of the following type are frequent:

$$\theta = \int_S \int_S g(\rho) d\mathbf{x} d\mathbf{z}, \tag{2.108}$$

where points $\mathbf{x}$ and $\mathbf{z}$ lie in the unit sphere S and ρ is the distance between $\mathbf{x}$ and $\mathbf{z}$; that is,

$$\rho = \sqrt{\sum_{i=1}^3 (x_i - z_i)^2}.$$

Consider the case where $\mathbf{x}$ and $\mathbf{z}$ are uniformly distributed throughout the sphere S. Let us first approach this problem in naive fashion. We embed the sphere in a cube of length 2 on a side. Then, we sample $x_j (j = 1, 2, 3)$ independently from $\mathcal{U}(-1, +1)$. If

$$||\mathbf{x}||^2 = \sum_{j=1}^{3} x_j^2 \leq 1,$$

then $\mathbf{x}$ lies in the unit sphere and may be used as a random observation from the unit distribution on the interior of the unit sphere. If $||\mathbf{x}||^2 > 1$, $\mathbf{x}$ lies outside the sphere and is discarded. This truncation procedure works very well in general provided that the dimension of the sphere is not too large.

However, we recall that the volume of a k-sphere of radius r is given by

$$V_s = \frac{\sqrt{\pi^k r^{2k}}}{\Gamma(k/2 + 1)},$$

whereas the volume of an n cube of side $2r$ is $V_c = 2^k r^k$. For $k = 3$, $V_s/V_c = .5236$. Consequently, less than half the points sampled in the cube will be discarded. However, for $k = 6$, $V_s/V_c = .0807$. And for $k = 10$, $V_S/V_C = .0025$.

Having obtained N $\mathbf{x}_i$'s and N $\mathbf{z}_i$'s as indicated above, we could then use

$$\hat{\theta} = \frac{(4\pi/3)^2}{N} \sum_{i=1}^{N} g(\rho_i) \qquad (2.109)$$

as an estimate for θ. Moreover, we have as an estimate for $\text{Var}(\hat{\theta})$,

$$s_{\hat{\theta}}^2 = \frac{1}{N(N-1)} \sum_{i=1}^{N} [(4\pi/3)^2 g(\rho_i) - \hat{\theta}]^2. \qquad (2.110)$$

An approximate 95% confidence interval for θ is given by

$$\hat{\theta} - 2s_{\hat{\theta}} < \theta < \hat{\theta} + 2s_{\hat{\theta}}. \qquad (2.111)$$

Next, let us consider a strategy which exploits the symmetry of the problem. We can without loss of generality consider what is happening inside a circular cross-section centered at the origin as shown in Figure 2.2.

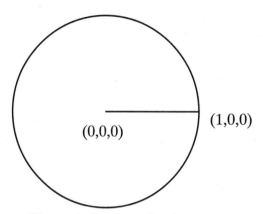

Figure 2.2. Cross Section.

Moreover, we can assume that $\mathbf{x}$ lies on the x_3 axis of the circle formed by slicing the sphere through the origin perpendicular to the x_2 axis as shown in Figure 2.3. Call x_3, τ_x.

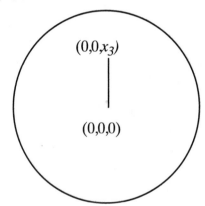

Figure 2.3. Slice Through Origin.

We cannot, however, assume that x_3 is sampled from $\mathcal{U}(0,1)$, as this would contradict the original assumption that $\mathbf{x}$ is uniformly distributed throughout the unit sphere. The density of $\mathbf{x}$ in Cartesian coordinates is

$$h(\mathbf{x}) \quad = \quad \frac{1}{4\pi/3} \text{ if } x_1^2 + x_2^2 + x_3^2 < 1 \tag{2.112}$$

$$= \quad 0 \quad \text{otherwise.}$$

In spherical coordinates the density is given by

$$f(\tau,\theta,\psi) = \frac{1}{4\pi/3}\sin(\psi)\tau^2 d\tau d\psi d\phi. \tag{2.113}$$

This gives as the marginal density of τ

$$f(\tau) \quad = \quad 3\tau^2 \text{ for } 0 < \tau < 1 \tag{2.114}$$

$$= \quad 0 \quad \text{otherwise.}$$

Next, we may take $\mathbf{z}$ to be in the x_1-x_3 plane. The point $\mathbf{z}$ has the joint cdf

$$F(\tau_z, \psi) = \frac{1}{2/3} \int_{-\pi}^{\psi} \int_0^{\tau_z} \tau^2 \sin \xi d\xi d\tau \qquad (2.115)$$

or

$$\begin{aligned} F(\tau_z) &= t_z^2 \text{ if } \tau_z < 1 \\ F(\psi) &= \frac{1}{2}[1 - \cos \psi] \text{ if } 0 < \psi < \pi. \end{aligned}$$

Thus

$$\begin{aligned} \rho &= \sqrt{(\tau_x - \tau_z \cos \psi)^2 + (\tau_z \sin \psi)^2} \\ &= \sqrt{\tau_x^2 + \tau_z^2 - 2\tau_x\tau_z \cos \psi}. \end{aligned}$$

Thus, we have reduced a six-dimensional problem to one of only three dimensions. The sampling of τ_x and τ_z is easily accomplished by sampling w from $\mathcal{U}(0, 1)$. Then

$$\begin{aligned} F(\tau) = \tau^3 &= w \\ \text{and } \tau &= w^{\frac{1}{3}}. \end{aligned}$$

The sampling of ξ is accomplished by taking v from $\mathcal{U}(0, 1)$. Then

$$\begin{aligned} F(\psi) = \frac{1}{2}[1 - \cos \psi] &= v \\ \cos(\psi) &= 1 - 2v \\ \psi &= \cos^{-1}(1 - 2v). \end{aligned}$$

Our estimator, then, is

$$\hat{\theta} = \frac{1}{N} \sum_{i=1}^{N} g(\psi(\tau_x, \tau_z, \psi)_i). \qquad (2.116)$$

Its estimated variance is

$$s_{\hat{\theta}}^2 = \frac{1}{N(N-1)} \sum_{i=1}^{N} [g(\rho_i) - \hat{\theta}]^2. \qquad (2.117)$$

2.10   Stratification in Multidimensional Integration

We return to the problem of evaluating $\theta = \int_0^1 g(\mathbf{x})d\mathbf{x}$ given that $\mathbf{x}$ is a k-dimensional vector. A naive quadrature rule might be to divide the unit hypercube into $N = n^k$ hypercubes. For $k = 2$, $n = 3$ the picture would be as shown in Figure 2.4:

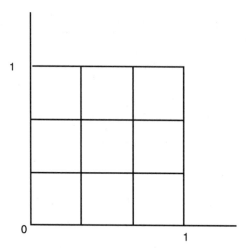

Figure 2.4. 2 × 2 Hypercube.

We might then evaluate $g(\mathbf{x})$ at the center of each subcube and take the average:

$$\theta \approx \frac{1}{9} \sum_{i=1}^{9} g(\mathbf{x}_i).$$

(2.118)

Or, in general, as shown in Figure 2.5,

$$\theta \approx \int_0^1 g(\mathbf{x})dx \approx \frac{1}{N} \sum_{i=1}^{N=n^k} g(\mathbf{x}_i).$$

(2.119)

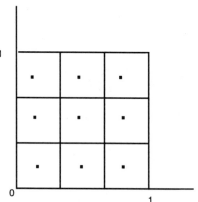

Figure 2.5. Equal Mesh.

The sample mean Monte Carlo analogue of this approach would be to sample a point from each subcube randomly and take the average:

$$\theta = \int_0^1 \int_0^1 g(x_1, x_2)dx_1 dx_2 \approx \frac{1}{9} \sum_{i=1}^{9} g(\mathbf{x}_i).$$

(2.120)

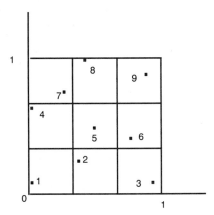

Figure 2.6. Stratified Monte Carlo.

Or, in general (Figure 2.6):

$$\theta = \int_0^1 g(\mathbf{x})d\mathbf{x} \approx \frac{1}{N}\sum_{i=1}^{N=n^k} g(\mathbf{x}_i) = \hat{\theta}. \qquad (2.121)$$

Clearly,

$$E(\hat{\theta}) = \theta. \qquad (2.122)$$

It can be shown that if $\partial g/\partial x_1$, $\partial g/\partial x_2$, ...,$\partial g/\partial x_k$, are all continuous and bounded in the hypercube, then $\mathrm{Var}(\hat{\theta})$ decreases like $N^{-(1+2/k)}$. Clearly, for k large, this strategy is not much to be preferred to the sample mean Monte Carlo approach without stratification. However, if we know enough about $g(\mathbf{x})$ to determine in which subcubes $g(\mathbf{x})$ is varying most, we could modify the strategy to sample more points where the variation is greater.

In one-dimensional stratification, we sampled a number of points from each stratum. In $k-$dimensional integration, this is also a possibility. For example, if we sample p points from each subcube, $\mathrm{Var}(\hat{\theta})$ will decrease like $1/(pN^{1+2/k})$. However, if k is large, n^k is likely to be so large that we will have to settle for p small.

We have noted that for k large, one is almost driven to Monte Carlo because of the impracticality of finding a reasonable grid for the quadrature procedure. For k that large, probably stratification is not the answer either. Stratification is, after all, a marriage between a regular grid procedure and "random quadrature."

However, it appears reasonable that in some cases, one might do as well to use a quasi Monte Carlo approach as we shall outline below. Recall that a principal advantage of blind stratification is to guarantee that our sampling procedure will have points more or less evenly demonstrated throughout the hypercube. We shall do so using the Halton sequence procedure [1, 4] discussed in Chapter 1.

We shall consider then, the Halton sequence, defined as a sequence of ordered k-tuples such that

$$\mathbf{W}_j = (p_{r_1}(j), p_{r_2}(j), \ldots, p_{r_k}(j)), \qquad (2.123)$$

where the k r_is are mutually prime.

Example. Suppose that $k = 3$, $r_1 = 2$, $r_2 = 3$, $r_3 = 5$, as shown in Table 2.8.

Table 2.8		
Halton Sequence		
i	$p_2(i)$	$p_3(i)$	p_5
1	1/2	1/3	1/5
2	1/4	2/3	2/5
3	3/4	1/9	3/5
4	1/8	4/9	4/5
5	5/8	7/9	1/25
6	3/8	2/9	6/25

Such sequences give the appearance of being randomly distributed throughout a unit hypercube. They are unlikely to be in synchrony with an integrand g. Unlike simpleminded sample mean Monte Carlo, a Halton sequence will not give for moderate sample size a clustering of points in subregions of the hypercube with few or no points in other subregions of equal volume.

We shall use as an estimator for $\theta = \int_0^1 g(\mathbf{x})d\mathbf{x}$,

$$\hat{\theta}_H = \frac{1}{N} \sum_{j=1}^N g(\mathbf{W}_j). \qquad (2.124)$$

Let g have the property that all partial derivatives containing not more than one differentiation with respect to each coordinate are piecewise continuous and bounded by a constant. It can be shown that $l.u.b.|\int_0^1 g(\mathbf{x})d\mathbf{x} - \hat{\theta}_H|$ decreases like $(\log(N))^k/N$.

In fact, if there exists a constant L such that

$$\left|\frac{\partial^m g(\mathbf{x})}{\partial x_{i_1} \partial x_{i_2} \ldots \partial x_{i_m}}\right| \leq L \qquad (2.125)$$

where $\{i_1, i_2, \ldots, i_m\}$ is any subset without repetition of $\{1, 2, \ldots, k\}$, then for fairly large N,

$$\left|\int_0^1 g(\mathbf{x})d\mathbf{x} - \frac{1}{N} \sum_{j=1}^N g(\mathbf{W_j})\right| \leq LB \frac{(\log N)^k}{N} \qquad (2.126)$$

where

$$B = \Pi_{i=1}^k \frac{r_i - 1}{\log r_i}.$$

Clearly, since r_i grows faster than $\log r_i$, it is advantageous to choose the $\{r_i\}$ to be small.

Example. We consider the evaluation of

$$\int_0^1 \int_0^1 \int_0^1 \exp[-x_1^2 - x_2^2 - x_3^2]dx_1 dx_2 dx_3 = \theta$$

Of course we could easily carry out the integration as a one-dimensional integral. However, we use a three-dimensional Halton sequence for reasons of demonstration in Table 2.9:

		Table 2.9		
		Halton Integration		
i	$p_2(i),p_3(i),p_5$	$\exp[-p_2^2(i) - p_3^2(i) - p_5^2(i)]$		
1	(.5,.3333,.2)	.8077
2	(.25,.6667,.4)	.5133
3	(.75,.1111,.6)	.3926
4	(.125,.4444,.8)	.4261
5	(.625,.7778,.04)	.3689
6	(.375,.2222,.024)	.8265
		3.3351

$$\hat{\theta} = \frac{1}{6}(3.33351) = .55585$$

Moreover,

$$L = g(0,0,0) = 1$$
$$B = \frac{2-1}{\log 2}\frac{3-1}{\log 3}\frac{5-1}{\log 5} = \frac{8}{\log 2 \log 3 \log 5} = 6.5275.$$

Thus, we should expect that for N large

$$|\theta - \hat{\theta}| \le 6.5275\frac{(\log N)^3}{N}.$$

2.11 Wiener Measure and Brownian Motion

Let us now consider the computation of integrals of the form

$$\int_{\mathcal{X}} F(x)d\mu_W(x), \qquad (2.127)$$

where $\mathcal{X}$ is the function space of all functions $x(t)$ which are continuous over the "time" interval $[0,T]$, and for which $x(0) = 0$; F is an arbitrary functional on $\mathcal{X}$; and $d\mu_W$ is Wiener measure. To define Wiener measure, let us consider the conditional probability distribution of $x(t + h)$. If at time t the value of x is given as $x(t)$, then the probability distribution of $x(t + h)$ $(h > 0)$ is $\mathcal{N}(x(t), h)$.

This description gives us an immediate simulation construction. First, we divide $[0, T]$ into the k intervals $[0, T/k]$, $[T/k, 2T/k]$, ..., $[(k-1)T/k, T]$. Then, if $\{\xi_j\}$, is a collection of independent normal random variates with mean 0 and variance 1, we let

$$
\begin{aligned}
x_1 &= \sqrt{\frac{T}{k}}\xi_1 \\
x_2 &= \sqrt{\frac{T}{k}}\xi_2 + x_1 \\
&\cdots \qquad \cdots \\
x_k &= \sqrt{\frac{T}{k}}\xi_k + x_{k-1}.
\end{aligned}
\tag{2.128}
$$

Then we may approximate (2.128) by $\hat{F}(x_1, x_2, \ldots, x_k)$, where $\hat{F}$ is the simpleminded discrete approximation to F.

Starting again from $t = 0$, we generate another random sample of size k:

$$
\begin{aligned}
x_{k+1} &= \sqrt{\frac{T}{k}}\xi_{k+1} \\
x_{k+2} &= \sqrt{\frac{T}{k}}\xi_{k+2} + x_{k+1} \\
&\cdots \qquad \cdots \\
x_{2k} &= \sqrt{\frac{T}{k}}\xi_{2k} + x_{2k-1}.
\end{aligned}
\tag{2.129}
$$

Then, after the generation of N polygonal paths,

$$
\int_{\mathcal{X}} F(x) d\mu_W(x) \approx \frac{1}{N} \sum_{j=0}^{N-1} \hat{F}(x_{jk+1}, \ldots, x_{(j+1)k}) = \tilde{\theta}_{k,N}.
\tag{2.130}
$$

An approximate value for the variance is given by:

$$
\tilde{V}(\tilde{\theta}_{k,N}) \approx \frac{1}{N} \sum_{j=0}^{N-1} [\hat{F}(x_{jk+1}, \ldots, x_{(j+1)k}) - \tilde{\theta}_{k,N}]^2.
\tag{2.131}
$$

Only in the limit as the spacing for the time grid goes to zero and the number of paths goes to infinity does $\tilde{\theta}_{k,N}$ equal (2.128). However, for reasonable F, it will be true that for k and N sufficiently large

$$
\tilde{\theta}_{k,N} \approx \int_{\mathcal{X}} F(x) d\mu_W(x).
\tag{2.132}
$$

A Brownian process stands in very much the same relation to a Wiener integral as a differential equation does to a definite integral. The analogue of the assumption of the knowledge of the functional F for the Wiener integral is the assumption that the right hand side equals $d\ln(X)$ (and of course, many other models could be considered).

Problems

2.1. Consider the evaluation of the integral

$$\int_0^1 |\sin 1000\pi x|dx.$$

(a) First approximate the integral by Simpson's Rule (2.2) using as mesh $0(.01)1$.
(b) Next, estimate the integral using sample mean Monte Carlo with a sample of size 100.
(c) Explain the difference in the two kinds of estimates.

2.2. Consider the evaluation of the integral

$$\int_1^3 \frac{\sqrt{x}}{\sqrt{x+1}} \exp(-x)dx.$$

(a) Use importance sampling to estimate the value of the integral.
(b) Use control variate sampling to estimate the value of the integral.

2.3. Consider the evaluation of the integral

$$\int_1^5 \frac{x^2}{x^2+5} \exp(x^2)dx.$$

Carry out the evaluation by dividing the interval as per $1(.1)5$ and using a stratified sampling strategy.

2.4. Consider the evaluation of the integral

$$\int\int\int_{25 \geq x^2+y^2+z^2 \geq 1} \frac{x^2+y^2+z^2}{x^2+y^2+z^2+5} \exp(x^2+y^2+z^2)dxdydz.$$

Here is an example where the use of Cartesian intervals is very inefficient. Show how a transformation to spherical coordinates, followed by stratification on the variable $r = \sqrt{x^2+y^2+z^2}$ works efficiently.

2.5. Consider again estimation of the integral

$$\int_1^5 \frac{x^2}{x^2+5} \exp(x^2)dx$$

Use the antithetic variables approach in (2.62) with five intervals to estimate the integral.

2.6. Let us now consider the computation of

$$\int_{\mathcal{X}} F(x)d\mu_W(x),$$

where $\mathcal{X}$ is the function space of all functions $x(t)$ which are continuous over the "time" interval $[0,T]$ and for which $x(0) = 0$, F is an arbitrary functional on $\mathcal{X}$, and $d\mu_W$ is Wiener measure.
(a) Estimate when $F = x$ and $T = 1$.
(b) Estimate when $F = x^2$ and $T = 1$.
(c) Estimate when $F = e^x$ and $T = 1$.

2.7. Consider the estimation of the integral (2.102)

$$\int_0^1 \int_0^1 \int_0^1 \int_0^1 \int_0^1 \int_0^1 \frac{\exp[\sqrt{x_1} + x_2 x_3 + x_4^2 x_5]}{\sqrt{x_1 x_4 + x_3 x_6 + .01}} \Pi_{j=1}^6 dx_j.$$

(a) Carry this out using Halton sequences of radices 2, 3, 5, 7, 11, and 13.
(b) Compare the result with that obtained by sample mean Monte Carlo.

References

[1] Halton, I.H. (1970). "A retrospective and prospective survey of the Monte Carlo method, *SIAM Rev.*, **12**, 1-63.

[2] Hammersley, J.M. and Handscomb, D.C. (1964). *Monte Carlo Methods.* New York: John Wiley & Sons, 60−74.

[3] Rubinstein, R.V. (1981). *Simulation and the Monte Carlo Method.* New York: John Wiley & Sons, 114−157.

[4] Shreider, Y.A. (1966). *The Monte Carlo Method: The Method of Statistical Trials.* New York: Pergamon Press, 131−134.

[5] Thompson, J.R. (1989). *Empirical Model Building.* New York: John Wiley & Sons, 107−108.

[6] Williams, E.E. and Thompson, J.R. (1998). *Entrepreneurship and Productivity.* New York: University Press of America, 235−239.

Chapter 3

Monte Carlo Solutions of Differential Equations

3.1   Introduction

John von Neumann, putative claimant to the title "founder of computer science," was motivated to conceive of and build the first serious digital computer as a device for handling simulation algorithms which he had formulated for dealing with problems in nuclear engineering.[1] Ideally, if we are dealing with problems of heat transfer, neutron flux, and so on, in regular and symmetrical regions, the classical nineteenth and early twentieth century differential-integral-difference equation formulations can be utilized. However, if the regions are complicated, if indeed we are concerned about a maze of pipes, cooling vessels, rods, and so on, the closed-form solutions are not available. This means that many person-years would be required to come up with all the approximation-theoretic quadrature calculations to ensure that a satisfactory plant will result if the plans are implemented. Von Neumann noticed that if large numbers of simple repetitive computations could be readily performed by machine, a method could be devised which would serve as an alternative to quadrature.

In reality, the quadrature issue, which Monte Carlo was largely developed to address, is rather unimportant compared to the much more important issue of direct simulation. To make a distinction between Monte Carlo and simulation, let us consider the following two paradigms shown in Figure 3.1. In the upper flowchart, we note a traditional means of coping with the numerical results of a model. We start out with axioms at the micro level which are generally easily understood. For example, one such axiom

---

[1]The discussion in this chapter largely follows [3].

might be that a gas particle starts at a particular point and moves step by step in three-space according to specified laws until it collides with a wall. Dealing with each specific gas molecule out of a total of, say, 10^{12} molecules is a hopeless task. Thus, investigators in the nineteenth century quite naturally and correctly were led to means for summary information about the gas molecules. That is to say, they had to content themselves with differential-integral-equation models as average representations of the effect of trillions of molecules.

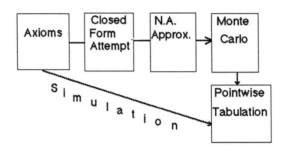

Figure 3.1. Two Ways of Problem Solving.

In Figure 3.1 the upper path gives the paradigm for solving such problems based on precomputer age models. We start with axioms which are accepted by most investigators in the field. These are transformed into a differential-integral-difference-equation type of summary model. Then, a generally *pro forma* attempt is made to arrive at a closed-form solution, that is, a representation which can be holistically comprehended by an observer and which lends itself to precise numerical evaluation of the dependent variables as we change the parameters of the model and the independent variables. This attempt is generally unsuccessful and leads only to some nonholistic quadrature-like setup for numerical evaluation of the independent variables. If the dimensionality of the quadrature is greater than 2, the user moves rather quickly to a random quadrature Monte Carlo approach.

What would have happened had computers been developed a century before they were? Would differential-integral equation modeling be the backbone of so much of physical science the way it still is today? It is an open question.

The fact is that we now have the computer speed to use the algorithm in the lower part of the diagram. We can now frequently dispense with the traditional approach by one which goes directly from the microaxioms to pointwise evaluation of the dependent variables. The technique for making this "great leap forward" is, in principle, simplicity itself.

Simulation carries out that which would earlier have been thought to be impossible, namely, to follow the progress of the particles, the cells, whatever. We do not do this for all the particles, but for a representative sample. We still do not have the computer speed to deal with 10^{10} particles; but we can readily deal with, say, 10^4 or 10^5. For many purposes, such a size is more than sufficient to yield acceptable accuracy. Among the advantages of a simulation approach is principally that it enables us to eliminate time-consuming and artificial approximation-theoretic activities and spend our time in more useful pursuits.

More importantly, simulation enables us to deal with problems which are so complex in their "closed-form" manifestation that they are presently attacked only in *ad hoc* fashion. For example, econometric approaches are generally linear, not because such approaches are supported by microeconomic theory, but because the complexities of dealing with the nonlinear consequences of the microeconomic theory are so overwhelming. Similarly, in mathematical oncology, the use of linear models is motivated by the failure of the natural branching process models to lead to numerically approximateable closed forms.

We have long since passed the point where computers can enable us to change fundamentally the ways we pose and solve problems. We have had the hardware capabilities for a long time to implement all the techniques covered in this chapter. But the proliferation of fast computing to the desktop will encourage private developers to develop simulation-based procedures for a large and growing market of users who need to get from specific problems to useful solutions in the shortest time possible. We now have the ability to use the computer, not as a fast calculator, but as a device which changes fundamentally the process of going from the microaxioms to the macrorealization.

3.2   Gambler's Ruin

There is an old temptation in applied mathematics to pose new problems, whenever possible, in classical "toy problem" formulation. One such is that of "gambler's ruin" [1]. We consider two gentlemen gamblers, A and B, who start to gamble in a zero-sum game with stakes x and $b - x$, respectively. At each round, each gambler puts up a stake of h dollars. The probability that A wins a round is p, while the probability that B wins a round is $q = 1 - p$. We wish to compute the probability that A ultimately wins the game. Let us define $v(x,t)$ to be the probability that A wins the game starting with capital x on or before the tth round. Similarly, $u(x,t)$ is the probability that B wins the game with his stake of $b - x$ on or before the tth round. Let $w(x,t)$ be the probability the game has not terminated by the tth round.

Each of the three variables v, u and w is bounded below by zero and above by one. Moreover, u and v are nondecreasing in t. w is nonincreasing in t. Thus, we can take limits of each of these as t goes to infinity. We shall call these limits $v(x)$, $u(x)$, and $w(x)$, respectively.

Although we shall briefly digress to get the closed-form solution to gambler's ruin, such a solution is really unimportant for our differential-integral simulation purposes. It will be the fundamental recursion in (3.1), which will be the basis for practically everything we do in this section.

$$v(x,t) = pv(x+h, t-\lambda) + qv(x-h, t-\lambda). \tag{3.1}$$

That is, the probability A wins the game on or before the tth round is given by the probability that he wins the first round and then ultimately wins the game with his new stake of $x+h$ in $t-\lambda$ rounds plus the probability he loses the first round and then wins the game in $t-\lambda$ rounds with his new stake of $x-h$. Here we have used a time increment of λ.

Taking limits in (3.1), we have

$$(p+q)v(x) = pv(x+h) + qv(x-h). \tag{3.2}$$

Rewriting (3.2), we have

$$p[v(x+h) - v(x)] = q[v(x) - v(x-h)]. \tag{3.3}$$

Let us make the further simplifying assumption that $b = Nh$. Then

$$v(\{n+1\}h) - v(nh) = (q/p)[v(nh) - v(\{n-1\}h)]. \tag{3.4}$$

We note that $v(0) = 0$ and $v(Nh) = 1$. Then writing (3.4) in extenso, we have

$$
\begin{aligned}
v(nh) - v(\{n-1\}h) &= (q/p)[v(\{n-1\}h) - v(\{n-2\}h)] \\
v(\{n-1\}h) - v(\{n-2\}h) &= (q/p)[v(\{n-2\}h) - v(\{n-3\}h)] \\
\cdots &= \cdots \\
\cdots &= \cdots \\
\cdots &= \cdots \\
v(2h) - v(h) &= q/p[v(h) - v(0)] = (q/p)v(h).
\end{aligned}
\tag{3.5}
$$

Substituting up the ladder, we have

$$v(nh) - v(\{n-1\}h) = (q/p)^{n-1}v(h). \tag{3.6}$$

Substituting (3.5) in the extenso version of (3.4), we have

$$
\begin{aligned}
v(nh) - v(\{n-1\}h) &= (q/p)^{n-1}v(h) \\
v(\{n-1\}h) - v(\{n-2\}h) &= (q/p)^{n-2}v(h) \\
\cdots &= \cdots \\
\cdots &= \cdots \\
\cdots &= \cdots \\
v(h) &= v(h).
\end{aligned}
\tag{3.7}
$$

Adding, we have

$$v(nh) = \left(1 + \frac{q}{p} + \left(\frac{q}{p}\right)^2 + \ldots + \left(\frac{q}{p}\right)^{n-1}\right)v(h). \tag{3.8}$$

Recalling that $v(Nh) = 1$, we have

$$1 = \left(1 + \frac{q}{p} + \left(\frac{q}{p}\right)^2 + \ldots + \left(\frac{q}{p}\right)^{N-1}\right)v(h) \tag{3.9}$$

Thus, we have

$$v(nh) = \frac{1 + q/p + (q/p)^2 + \ldots + (q/p)^{n-1}}{1 + q/p + (q/p)^2 + \ldots + (q/p)^{N-1}} \tag{3.10}$$

For $p = q = .5$, this gives,

$$v(nh) = \frac{n}{N} \ ; \ \text{i.e.,} \ v(x) = \frac{x}{b}. \tag{3.11}$$

Otherwise, multiplying (3.10) by $[1 - p/q]/[1 - p/q]$, we have

$$v(nh) = \frac{1 - (q/p)^n}{1 - (q/p)^N}, \quad \text{i.e.,} \tag{3.12}$$

$$v(x) = \frac{1 - (q/p)^{x/h}}{1 - (q/p)^{b/h}}.$$

Now, by symmetry,

$$u(x) = \frac{(q/x)^{x/h} - (q/h)^{b/h}}{1 - (q/p)^{b/h}} \tag{3.13}$$

From (3.12) and (3.13), we have

$$v(x) + u(x) = 1. \tag{3.14}$$

Consequently, $w(x) = 0$; that is, the game must terminate with probability 1. Thus, we can use a simulation to come up with reasonable estimates of the probability of A ultimately winning the game. A flowchart of such a simulation is given in Figure 3.2.

We note that this simulation gives us a ready means of estimating a rough 95% confidence interval for $v(x)$, namely

$$v(x) = \frac{W}{M} \pm \frac{2\sqrt{W(1 - W/M)}}{M}. \tag{3.15}$$

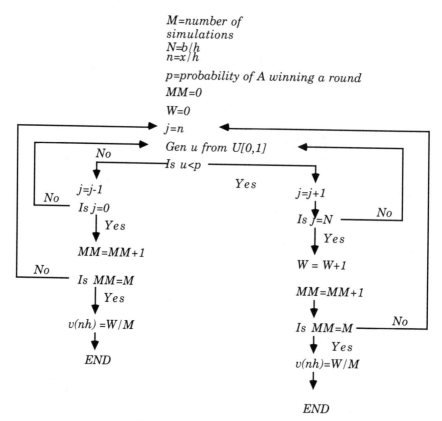

Figure 3.2. Gambler's Ruin.

3.3   Solution of Simple Differential Equations

Since we have shown a closed-form solution for the gambler's ruin problem, it would be ridiculous for us to use a simulation to solve it. It is by means of an analogy of real-world problems to the general equation (3.1) that simulation becomes useful. Rewriting (3.3), we have

$$p\Delta v(x) = q\Delta v(x - h),   \tag{3.16}$$

where

$$\Delta v(x) = [v(x + h) - v(x)]/h.$$

Subtracting $q\Delta v(x)$ from both sides of (3.16), we have

$$(p - q)\Delta v(x) = q[\Delta v(x - h) - \Delta v(x)],   \tag{3.17}$$

or

$$\Delta^2 v(x) + \frac{p - q}{qh}\Delta v(x) = 0,   \tag{3.18}$$

where

$$\Delta^2 v(x) = \frac{\Delta v(x) - \Delta v(x - h)}{h}.$$

For h sufficiently small, this is an approximation to

$$\frac{d^2 v}{dx^2} + 2\beta \frac{dv}{dx} = 0, \tag{3.19}$$

where

$$\frac{p - q}{qh} = 2\beta.$$

Now, suppose that we are given the boundary conditions of (3.19), $v(0) = 0$ and $v(b) = 1$. Then our flowchart in Figure 3.2 gives us a ready means of approximating the solution to (3.19). We simply set $p = (2\beta h+1)/(2\beta h+2)$, taking care to see that h is sufficiently small, if β be negative, to have p positive. To make sure that we have chosen h sufficiently small that the simulation is a good approximation to the differential equation, typically we use simulations with successively smaller h until we see little change in $v(x) \approx W/N$.

Suppose that the boundary conditions are less accommodating, for example, suppose that $v(0)$ and $v(b)$ take arbitrary values. A moment's reflection shows that

$$v(x) \approx \frac{W}{N} v(b) + \left(1 - \frac{W}{N}\right) v(0). \tag{3.20}$$

Since a closed-form solution of (3.19) is readily available. We need not consider simulation for this particular problem. But suppose that we generalize (3.19) to the case where β depends on x:

$$\frac{d^2 v}{dx^2} + 2\beta(x) \frac{dv}{dx} = 0. \tag{3.21}$$

Again, we use our flowchart in Figure 3.2, except that at each step we change p via

$$p(x) = \frac{2\beta(x)h + 1}{2\beta(x)h + 2}. \tag{3.22}$$

Once again, $v(x) \approx W/N v(b) + (1 - W/N)v(0)$. And once again, it is an easy matter to come up with an internal measure of accuracy via (3.15).

It is possible to effect numerous computational efficiencies. For example, we need not start afresh for each new grid value of x. For each pass through the flowchart, we can note all grid points visited during the pass and increase the counter of wins at each of these if the pass terminates at b, the number of losses if the pass terminates at 0.

3.4   Solution of the Fokker–Planck Equation

It is important to note that the simulation used to solve (3.19) actually corresponds, in many cases, to the microaxioms of which the differential equation (3.19) is a summary. This is very much the case for the Fokker-Planck equation which we consider below. Let us suppose that we do not eliminate time in (3.1). We will define

$v(x, t, 0; h, \lambda) = P[$particle starting at x will be absorbed at 0 on or before $t = m\lambda]$;

$v(x, t, b; h, \lambda) = P[$particle starting at x will be absorbed at b on or before $t = m\lambda$] ;

$V(x, t; h, \lambda) = V(0, t)v(x, t, 0; h, \lambda) + V(b, t)v(x, t, b; h, \lambda).$

We define

$$\Delta_t V(x, t; h, \lambda) = \frac{V(x, t + \lambda; h, \lambda) - V(x, t; h, \lambda)}{\lambda}$$

$$\Delta_x V(x, t; h, \lambda) = \frac{V(x + h, t; h, \lambda) - V(x, t; h, \lambda)}{h}$$

$$\Delta_{xx} V(x, t; h, \lambda) = \frac{\Delta_x V(x, t; h, \lambda) - \Delta_x V(x - h, t; h, \lambda)}{h}.$$

That is, the expected payoff is given by the probability a particle is absorbed at the left at time t, multiplied by the boundary award $V(0, t)$ plus the probability the particle is absorbed at the right at time t times the boundary award $V(b, t)$.

Now, our basic relation in (3.1) still holds, so we have

$$V(x, t + \lambda; h, \lambda) = p(x)V(x + h, t; h, \lambda) + q(x)V(x - h, t; h, \lambda). \quad (3.23)$$

Subtracting $V(x, t; h, l)$ from both sides of (3.23), we have

$$
\begin{aligned}
\lambda\Delta_t V(x, t; h, l) &= p(x)[V(x + h, t; h, \lambda) - V(x, t; h, \lambda)] \quad (3.24)\\
&\quad + q(x)[V(x - h, t; h, \lambda) - V(x, t; h, \lambda)]\\
&= p(x)[V(x + h, t; h, \lambda) - V(x, t; h, \lambda)]\\
&\quad - q(x)[V(x, t; h, \lambda) - V(x - h, t; h, \lambda)]\\
&= hp(x)\Delta_x V(x, t; h, \lambda) - hq(x)\Delta_x V(x - h, t; h, \lambda)\\
&= h[p(x) - q(x)]\Delta_x V(x, t; h, \lambda)\\
&\quad + hq(x)[\Delta_x V(x, t; h, \lambda) - \Delta_x V(x - h, t; h, \lambda)]\\
&= h[p(x) - q(x)]\Delta_x V(x, t; h, \lambda) + h^2 q(x)\Delta_{x,x} V(x, t; h, \lambda).
\end{aligned}
$$

Letting $p(x) = [\beta(x) + 2h\alpha(x)]/[2\beta(x) + 2h\alpha(x)]$ and $q(x) = 1 - p(x)$, we have:

$$\lambda\Delta_t V(x, t; h, \lambda) = 2h^2\frac{\alpha(x)}{2\beta(x) + 2h\alpha x} + h^2\frac{\beta(x)}{2\beta(x) + 2h\alpha(x)}\Delta_{xx} V(x, t; h, \lambda).$$

$$(3.25)$$

Next, taking h very small with $\lambda/h^2 = \mu$, we have

$$\mu\Delta_t V(x,t;h,\lambda) = \frac{\alpha(x)}{\beta(x)}\Delta_x V(x,t;h,\lambda) + \frac{1}{2}\Delta_{xx}V(x,t;h,\lambda). \qquad (3.26)$$

So the simulation, which proceeds directly from the microaxioms, yields in the limit as the infinitesimals go to zero a practical pointwise evaluator of the usual Fokker–Planck equation:

$$2\mu\frac{\partial V}{\partial t} = 2\frac{\alpha(x)}{\beta(x)}\frac{\partial V}{\partial x} + \frac{\partial^2 V}{\partial x^2}. \qquad (3.27)$$

The Fokker–Planck equation is generally not solvable in closed form. Note that we have given a simulation-based approach for solving (3.27), but more importantly we have given a practical means for arriving at the consequences of the original axioms which brought about Fokker–Planck in the first place. So we again raise the intriguing possibility that had computers been available 100 years ago, Fokker and Planck might have simply represented their model in microaxiomatic format instead of giving a differential equation summary thereof. Again, our algorithm is essentially the flowchart in Figure 3.2 with a time counter added on.

3.5   The Dirichlet Problem

Next, we consider another common differential equation model of physics, that of Dirichlet. In $\mathcal{R}_k$, let there be given a bounded connected region S Γ (Figure 3.3). Let there be given a function $F(x)$ satisfying the equation of Laplace inside S :

$$\sum_{j=1}^{k}\frac{\partial^2\phi_j}{\partial x_j^2} = 0. \qquad (3.28)$$

The values of ϕ are given explicitly at every boundary point by the piecewise continuous function $f(Q)$, that is,

$$\phi(x)|_\Gamma = f(Q). \qquad (3.29)$$

For most boundaries and boundary functions, the determination of ϕ analytically is not known. The usual numerical approximation approach can require a fair amount of setup work, particularly if the dimensionality is 3 or greater. We exhibit below a simulation technique which is, in fact, an actualization of the microaxioms which frequently give rise to (3.28). Although our discussion is limited to $\mathcal{R}_2$, the generalization to $\mathcal{R}_k$ is quite straightforward. Let us superimpose over S a square grid of length h on a side. The points of intersection inside S nearest Γ will be referred to as *boundary nodes*. All other nodes inside S shall be referred to as *internal nodes*.

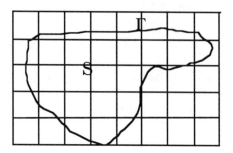

Figure 3.3. The Dirichlet Problem.

In Figure 3.4, we consider an internal node with coordinates (x, y) in relation to its four immediate neighbors.

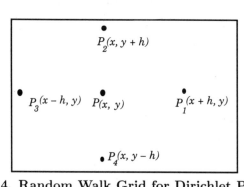

Figure 3.4. Random Walk Grid for Dirichlet Problem.

Now

$$\partial\phi/\partial x|_{x,y} \approx \frac{\phi(x + h/2, y) - \phi(x - h/2, y)}{h} \tag{3.30}$$

and

$$\frac{\partial^2 \phi}{\partial x^2} \approx \frac{(\partial\phi/\partial x)|_{x+h/2,y} - (\partial\phi/\partial x)|_{x-h/2,y}}{h} \tag{3.31}$$

$$\approx \frac{\phi(x + h, y) + \phi(x - h, y) - 2\phi(x, y)}{h^2}$$

Similarly,

$$\frac{\partial^2 \phi}{\partial y^2} \approx \frac{\phi(x, y + h) + \phi(x, y - h) - 2\phi(x, y)}{h^2}. \tag{3.32}$$

Equation (3.28) then gives

$$0 = \frac{\partial^2 \phi}{\partial x^2} + \frac{\partial^2 \phi}{\partial y^2} \approx \frac{\phi(P_1) + \phi(P_2) + \phi(P_3) + \phi(P_4) - 4\phi(P)}{h^2}. \tag{3.33}$$

So

$$\phi(P) \approx \frac{\phi(P_1) + \phi(P_2) + \phi(P_3) + \phi(P_4)}{4}. \tag{3.34}$$

Equation (3.34) gives us a ready means of a simulation solution to the Dirichlet problem. Starting at the internal node (x, y), we randomly walk to one of the four adjacent points with equal probabilities. We continue the process until we reach a boundary node, say Q_i. After N walks to the boundary from starting point (x, y), our estimate of $\phi(x, y)$ is given simply by

$$\phi(x, y) \approx \frac{\sum_{i=1}^{N} n_i f(Q_i)}{N}, \tag{3.35}$$

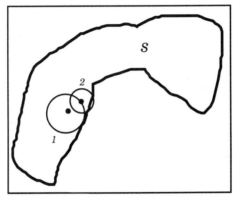

Figure 3.5. Quick Steps to the Boundary.

where n_i is the number of walks terminating at boundary node Q_i and the summation is taken over all boundary nodes.

In Figure 3.3, if we wish to show ϕ contours throughout S, we can take advantage of a number of computational efficiencies. For example, as we walk from (x, y) to the boundary, we will traverse $(x+h, y)$ numerous times. By incorporating the walks which traverse $(x + h, y)$ even though $(x + h, y)$ is not our starting point, we can increase the total number of walks used in the evaluation of $\phi(x + h, y)$.

Let us now consider in Figure 3.5 a technique which is particularly useful if we need to evaluate ϕ at only one point in S. Since it can easily be shown that $\phi(x, y)$, the solution to Laplace's equation inside S, is equal to the average of all values taken on a circle centered at (x, y) and lying inside S, we can draw around (x, y) the largest circle lying inside S and then select a point uniformly on the boundary of that circle, use it as the center of a new circle, and select a point at random on that circle. We continue the process until we arrive at a boundary node. Then, again after N walks,

$$\phi(x, y) \approx \frac{\sum_{i=1}^{N} n_i f(Q_i)}{N}. \tag{3.36}$$

The above method works, using hyperspheres, for any dimension k. Again, it should be emphasized that in many cases the simulation is a direct implementation of the microaxioms which gave rise to Laplace's equation.

3.6 Solution of General Elliptic Differential Equations

Next, we consider a general elliptic differential equation in a region S in two-dimensional space. Again, the values on the boundary Γ are given by $f(\cdot)$, which is piecewise continuous on Γ. Inside S,

$$\beta_{11}\frac{\partial^2\phi}{\partial x^2} + 2\beta_{12}\frac{\partial^2\phi}{\partial x\partial y} + \beta_{22}\frac{\partial^2\phi}{\partial y^2} + 2\alpha_1\frac{\partial\phi}{\partial x} + 2\alpha_2\frac{\partial\phi}{\partial y} = 0, \qquad (3.37)$$

where $\beta_{11} > 0$, $\beta_{22} > 0$, and $\beta_{11}\beta_{22} - \beta_{12}^2 > 0$. We consider the difference equation corresponding to (3.37), namely:

$$\beta_{11}\Delta_{xx}\phi + 2\beta_{12}\Delta_{xy}\phi + \beta_{22}\Delta_{yy}\phi + 2\alpha_1\Delta_x\phi + 2\alpha_2\Delta_y\phi = 0. \qquad (3.38)$$

As convenient approximations to the finite differences, we use

$$
\begin{aligned}
\Delta_{xy} &= \frac{\phi(x+h,y+h) - \phi(x,y+h) - \phi(x+h,y) + \phi(x,y)}{h^2} \\[1mm]
\Delta_{xx} &= \frac{\phi(x+h,y) + \phi(x-h,y) - 2\phi(x,y)}{h^2} \\[1mm]
\Delta_{yy} &= \frac{\phi(x,y+h) + \phi(x,y-h) - 2\phi(x,y)}{h^2} \\[1mm]
\Delta_x &= \frac{\phi(x+h,y) - \phi(x,y)}{h} \\[1mm]
\Delta_y &= \frac{\phi(x,y+h) - \phi(x,y)}{h}.
\end{aligned}
\qquad (3.39)
$$

These differences involve five points around (x,y). Now, we shall develop a random walk realization of (3.37), which we write out explicitly below in Figure 3.6:

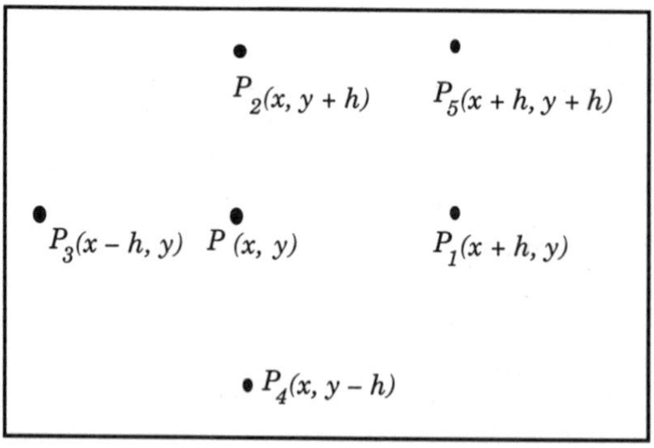

Figure 3.6. Grid for Elliptic Equation Random Walk.

$$\beta_{11} \frac{\phi(x+h,y) + \phi(x-h) - 2\phi(x,y)}{h^2}$$

$$+2\beta_{12} \frac{\phi(x+h,y+h) - \phi(x,y+h) - \phi(x+h,y) + \phi(x,y)}{h^2}$$

$$+\beta_{22} \frac{\phi(x,y+h) + \phi(x,y-h) - 2\phi(x,y)}{h^2}$$

$$+2\alpha_1 \frac{\phi(x+h,y) - \phi(x,y)}{h} + 2\alpha_2 \frac{\phi(x,y+h) - \phi(x,y)}{h} = 0. \quad (3.40)$$

We rearrange the terms in (3.40) to give

$$\phi(x+h,y)(\beta_{11} + 2\alpha_1 h - 2\beta_{12}) + \phi(x,y+h)(\beta_{22} + 2\alpha_2 h - 2\beta_{12})$$

$$+\phi(x-h,y)\beta_{11} + \phi(x,y-h)\beta_{22} + \phi(x+h,y+h)2\beta_{12}$$

$$= \phi(x,y)[2\beta_{11} - 2\beta_{12} + 2\beta_{22} + 2(\alpha_1 + \alpha_2 h)]. \quad (3.41)$$

Letting $D = [2\beta_{11} - 2\beta_{12} + 2\beta_{22} + 2(\alpha_1 + \alpha_2)h]$, we have

$$\phi(x+h,y)p_1 + \phi(x,y+h)p_2 + \phi(x-h,y)p_3 + \phi(x,y-h)p_4$$

$$+\phi(x+h,y+h)p_5 = \phi(x,y), \quad (3.42)$$

with

$$p_1 = \frac{\beta_{11} + 2\alpha_1 h - 2\beta_{12}}{D}; \quad p_2 = \frac{\beta_{22} + 2\alpha_2 h - 2\beta_{12}}{D};$$

$$p_3 = \frac{\beta_{11}}{D}; p_4 = \frac{2\beta_{22}}{D}; \quad p_5 = \frac{2\beta_{12}}{D}. \quad (3.43)$$

Note that in the formulation above , we must exercise some care to assure that the probabilities are nonnegative. By using the indicated probabilities, we walk randomly to the boundary repeatedly and use the estimate

$$\phi(x,y) = \frac{\sum_{i=1}^{N} n_i f(Q_i)}{N}. \quad (3.44)$$

3.7 Conclusions

The above examples are given to give the reader a feel as to the practical implementation of simulation-based algorithms as alternatives to the usual numerical approximation techniques. A certain amount of practice quickly brings the user to a point where he or she can write simulation algorithms in days to problems which would require the numerical analyst months to approach.

Other advantages of the simulation approach could be given. For example, since walks to boundaries are so simple to execute, it is easy to conceptualize the utilization of parallel processors to speed up the computations with a minimum of handshaking between the CPUs. But the main advantage of simulation is its ability to enable the user to bypass the traditional path in Figure 3.1 and go directly from the microaxioms to their macro consequences. Our algorithm for solving the Fokker–Planck problem and our algorithm for solving the Dirichlet problem are not simply analogues of the classical differential-equation formulations of these systems. *They are, in fact, descriptions of the axioms that typically give rise to these problems.* Here we note that the classical differential-equation formulation of many of the systems of physics and chemistry proceed from the axioms which form the simulation algorithm. It was simply the case, in a precomputer age, that the differential-integral equation formulation appeared to give the best hope of approximate evaluation via series expansions and the like.

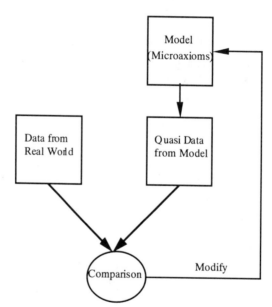

Figure 3.7. The Idealized Simulation Paradigm.

Moving further along, we now have the possibility of implementing the algorithm in Figure 3.7 as a basic model-building paradigm. In the path shown, computer simulation may obviate the necessity of modeling via, say, differential equations and then using the computer as a means of approximating solutions, either by numerical techniques or by Monte Carlo to the axiomitized model. The computer then ceases to be a fast calculator, and shifts into actually simulating the process under consideration. Such a path represents a real paradigm shift and will begin the realization of the computer as a key part of the scientific method.

Problems

3.1. Consider the differential equation

$$\frac{d^2v}{dx^2} + 2x^{2.5}\frac{dv}{dx} = 0,$$

where $v(0) = 1$ and $v(1) = 2$. Use the flowchart in Figure 3.2 to obtain estimates of $v(.2)$, $v(.4)$, $v(.6)$, and $v(.8)$, together with error bounds for these quantities.

3.2. Program a quadrature-type differential equation solver for the following differential equation, where $x(0) = 0$ and $x(1) = 1$:

$$\frac{d^2x}{dt^2} + \beta(t)\frac{dx}{dt} = 0.$$

Compare its performance with a simulation-based approach using Figure 3.2 for the following candidates for β:
(a) $\beta = 4$
(b) $\beta = t$
(c) $\beta = \sin(4\pi t/t + 1)$.

3.3. Consider the differential equation on the unit x interval.

$$\frac{\partial V}{\partial t} = 2\frac{\partial V}{\partial x} + \frac{\partial^2 V}{\partial x^2},$$

where $V(x,t) = 200x\exp(-t)$. Again, use the flowchart in Figure 3.2 to obtain estimates of $V(.2,t)$, $V(.4,t)$, $V(.6.t)$, and $V(.8,t)$, for $t = 0,1,2,3$.

3.4. Draw contours for T in the interior of the plate shown in Figure 3.8,

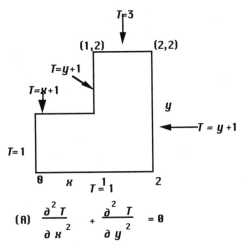

Figure 3.8. Dirichlet Problem for a Plate.

using increments of 0.2 in both x and y based on the Gambler's Ruin flowchart in Figure 3.2, given that Laplace's equation (A) is satisfied in the interior with the boundary conditions shown. Here, we have a substantial demonstration of how inefficient it would be to run simulation from each interior point (x, y) independently of walks from other points which "step on" (x, y).

3.5. Consider the equation

$$(1+y)\frac{\partial^2 \phi}{\partial x^2} + 2x\frac{\partial^2 \phi}{\partial x \partial y} + (1-y)\frac{\partial^2 \phi}{\partial y^2} - \frac{\partial \phi}{\partial x} = 0,$$

which is satisfied inside the unit circle shown in Figure 3.9, with the boundary condition indicated. Use the standard gambler's ruin flowchart in Figure 3.2 to obtain estimates of the ϕ contours inside the unit circle.

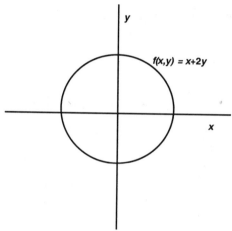

Figure 3.9. Elliptic Equation on the Circle.

References

[1] Lattes, R. (1969). *Methods of Resolution for Selected Boundary Problems in Mathematical Physics.* New York: Gordon and Breach.

[2] Shreider, Y.A. (1966). *The Monte Carlo Method: The Method of Statistical Trials.* New York: Pergamon Press.

[3] Thompson, J.R. (1989). *Empirical Model Building.* New York: John Wiley & Sons, 95–108.

Chapter 4

Markov Chains, Poisson Processes, and Linear Equations

4.1 Discrete Markov Modeling

4.1.1 The Basic Model

It is natural, in a number of situations, to use categorical descriptions of situations. For example, a layperson might have an implicit model concerning his or her oral temperature:

Temp. < 92.5	dangerously low
$92.5 \leq$ Temp. < 95.5	very low
$95.5 \leq$ Temp. < 97.5	low
$97.5 \leq$ Temp. < 98.8	normal
$98.8 \leq$ Temp. < 100.5	high
$100.5 \leq$ Temp. < 103.5	very high
$103.5 \leq$ Temp. < 107.0	dangerously high
$107.0 \leq$ Temp.	death

Suppose that a person examines his or her temperature daily. The temperature case is an example where a continuous variable is essentially discretized for decision purposes. For example if one is sick, at what point should one go the expense of a visit to a physician? The example above is one where we have eight states. A person might be expected to stay mostly in the normal range, with changes to other states as illnesses of one sort or another occur. There will be a natural migration to the normal state for some average person who enjoys reasonable health. On the other hand, we can well imagine that the probability that a person moves into the normal state from the very high state is a function of how long he or she has been in the very high state. But for some purposes, we might regard that the state one will be in tomorrow will only depend on the state he or she is in today. This one-step memory is called the *Markov property*. We notice that the death state is "absorbing," for there is no progression from the death state to any other.

Proceeding more abstractly, let us suppose that $E_1, E_2, \ldots, E_k$ are k states. Let us discretize time. Let there be random variable X_n, which is equal to i if we are in state E_i at time n. We now add a few definitions.

The probability of being in state j at time n is given by $P[X_n = j] = p_j^{(n)}$. Now, if $P[X_n = j | X_{n-1} = i] = p_{ij}^n$, irrespective of the history prior to $n-1$, the stochastic process is said to have the *Markov property* and is called a *Markov chain*. If the number of states is finite, we have a *finite Markov chain*. Suppose that $p_{ij}^n = p_{ij}$, that is, is independent of n, then the Markov chain is *homogeneous*. Clearly, then, $\sum_j p_{ij} = 1$.

The *transition matrix* of a Markov chain is given by

$$\mathbf{P} = \{p_{ij}\}^T = \begin{bmatrix} p_{11} & p_{21} & \cdots & p_{k1} \\ p_{12} & p_{22} & \cdots & p_{k2} \\ \vdots & \vdots & & \vdots \\ p_{1k} & p_{2k} & \cdots & p_{kk} \end{bmatrix}. \tag{4.1}$$

Now if

$$\mathbf{p} = \begin{bmatrix} p_1^{(n)} \\ p_2^{(n)} \\ \vdots \\ p_k^{(n)} \end{bmatrix},$$

then

$$\mathbf{p}^{(n)} = \mathbf{P}^n \mathbf{p}^0 = \mathbf{P}^{n-l} \mathbf{p}^l \text{ for } 0 \leq l \leq n. \tag{4.2}$$

A favorite example of the use of Markov chains is taken from Mendelian genetics. Suppose we have two genes **A** and **a**. This gives us three possible genotypes: **AA**, **Aa**, and **aa**. We mate two individuals and select for the next generation two other offspring of opposite sexes. The possible states are

$E_1 = \mathbf{AA} \times \mathbf{AA}$ $E_3 = \mathbf{Aa} \times \mathbf{Aa}$ $E_5 = \mathbf{aa} \times \mathbf{aa}$

$E_2 = \mathbf{AA} \times \mathbf{Aa}$ $E_4 = \mathbf{Aa} \times \mathbf{aa}$ $E_6 = \mathbf{AA} \times \mathbf{aa}$.

The transition matrix is given by

$$\mathbf{P} = \{p_{ij}\}^T = \begin{bmatrix} p_{11} & p_{21} & \cdots & p_{61} \\ p_{12} & p_{22} & \cdots & p_{62} \\ \vdots & \vdots & & \vdots \\ p_{16} & p_{26} & \cdots & p_{66} \end{bmatrix} = \begin{bmatrix} 1 & \frac{1}{4} & \frac{1}{16} & 0 & 0 & 0 \\ 0 & \frac{1}{2} & \frac{1}{4} & 0 & 0 & 0 \\ 0 & \frac{1}{4} & \frac{1}{4} & \frac{1}{4} & 0 & 1 \\ 0 & 0 & \frac{1}{16} & \frac{1}{4} & 1 & 0 \\ 0 & 0 & \frac{1}{8} & 0 & 0 & 0 \end{bmatrix}.$$

(4.3)

Now a little multiplication quickly reveals that

$$\mathbf{P}^8 = \begin{bmatrix} 1.000 & 0.671 & 0.408 & 0.173 & 0 & 0.383 \\ 0 & 0.050 & 0.059 & 0.046 & 0 & 0.073 \\ 0 & 0.059 & 0.073 & 0.059 & 0 & 0.091 \\ 0 & 0.046 & 0.059 & 0.050 & 0 & 0.073 \\ 0 & 0.173 & 0.408 & 0.671 & 1.000 & 0.383 \\ 0 & 0.009 & 0.011 & 0.009 & 0 & 0.014 \end{bmatrix}$$

(4.4)

and the limiting concatenated transition matrix is seen to be

$$\mathbf{P}^\infty = \begin{bmatrix} 1.000 & 0.750 & 0.500 & 0.250 & 0 & 0.500 \\ 0 & 0.000 & 0.000 & 0.000 & 0 & 0.000 \\ 0 & 0.000 & 0.000 & 0.000 & 0 & 0.000 \\ 0 & 0.000 & 0.000 & 0.000 & 0 & 0.000 \\ 0 & 0.250 & 0.500 & 0.750 & 1.000 & 0.500 \\ 0 & 0.000 & 0.000 & 0.000 & 0 & 0.000 \end{bmatrix}.$$

(4.5)

For set-piece problems such as Mendelian genetics, the Markov chain−based algorithms may be satisfactory, just as ordinary linear differential equations may well be suited for many problems in, say, mechanics. But it is easy to see situations where other approaches are better.

4.1.2 Saving the King

Let us consider the following hypothetical example from the realm of geopolitics. An intelligence analyst is attempting a study of the prognosis of a new ruler in a friendly Middle Eastern kingdom. The king of that country is involved in a struggle to withstand a takeover by Islamic radicals. At the moment, the king is a "favorable" **4** position. Other positions are "disaster" **1**, "unfavorable" **2**, "moderate" **3**, "strongly favorable" **5**, and "as secure as it gets in the Middle East" **6**. Based on all data, including the long track record of the new king's father, the following matrix of transition probabilities is arrived at (where the time frame is three months).

$$
P = \begin{bmatrix}
 & 1 & 2 & 3 & 4 & 5 & 6 \\
1 & 0.20 & 0.05 & 0.05 & 0.05 & 0.00 \\
0 & 0.20 & 0.05 & 0.05 & 0.05 & 0.00 \\
0 & 0.30 & 0.60 & 0.20 & 0.10 & 0.00 \\
0 & 0.15 & 0.10 & 0.50 & 0.10 & 0.00 \\
0 & 0.10 & 0.10 & 0.10 & 0.40 & 0.02 \\
0 & 0.05 & 0.10 & 0.10 & 0.30 & 0.98
\end{bmatrix}.
$$

Using this information, we can look at the profile of probabilities five years out by computing

$$
P^{20} = \begin{bmatrix}
1.00 & 0.4774 & 0.3445 & 0.3445 & 0.2644 & 0.0647 \\
0 & 0.0029 & 0.0036 & 0.0036 & 0.0036 & 0.0040 \\
0 & 0.0141 & 0.0174 & 0.0174 & 0.0168 & 0.0175 \\
0 & 0.0089 & 0.0110 & 0.0110 & 0.0108 & 0.0117 \\
0 & 0.0208 & 0.0260 & 0.0260 & 0.0283 & 0.0347 \\
0 & 0.4759 & 0.5976 & 0.5976 & 0.6760 & 0.8674
\end{bmatrix}.
$$

However, the situation is complicated by non-Markovian factors (that is, factors with a memory past one step). Two quarters together in either the unfavorable state or the strongly unfavorable state give a score of -3. The analyst opines that as soon as a score of -10 is reached, the king will flee the country. He wants to compute the probability that the king will still be in power in one year, in two years, in five years.

The simulation solution of this problem is straightforward. We simply walk through the postulated paths many times and use the resulting expectations. Essentially, we may analogize this to an urn sampling. We have six urns. We start sampling in the strongly favorable urn, namely 5. We sample from balls in that urn with relative probabilities (0.05, 0.05, 0.10, 0.10, 0.40, 0.30). Then we go to urn sampled, and so on, until the number of quarters, say M, has been achieved. We recall the special side condition. So, as soon as we land in urn 2 or 3, we add one to the BadQuarterCounter. When we move to any other urn, we reset the BadQuarterIndicator to zero. When we find the BadQuarterCounter to be 2 or greater, we subtract 3 from the BadQuarterPenalty. We keep a running count of the number of times we have landed in each of the urns (when we reach the absorbing state 1, we increase the accumulated 1 counter by one; whenever we reach -10 on the BadQuarterPenalty, we add one to the "flee the country counter"—say 7). So for a time span of 20 quarters, we have the relative numbers of times we land in each of the seven states, and we can give the analyst the kind of information he seeks. It is almost always the case, also, that we can give the analyst lots more information than that which he immediately seeks. For example, we can find the probability of flight before ever reaching the relatively secure state 6. The paradigm is straightforward. We start one simple program structure and leave the coding to the reader (Problem 4.1).

Input M = number of quarters to be simulated
Input N = number of simulated paths of length M
Input BadQuarterIndicator = 0
Input BadQuarterPenalty = 0
Input CumRuns = 0
Input Cum(j); j=1,6 = 0
Input cumulative interval probabilities as
1 [1.00, 1.00, 1.00, 1.00, 1.00, 1.00]
2 [0.20, 0.40, 0.70,0.85, 0.95. 1.00]
3 [0.05, 0.10, 0.70, 0.80, 0.90, 1.00]
4 [0.05, 0.10, 0.20, 0.30, 0.70, 1.00]
5 [0.05, 0.10, 0.20 , 0.30,0 .70, 1.00]
6 [0.00, 0.00, 0.00,0.00, 0.02, 1.00]

Since we start in stage **5**, we begin there

5 CumRuns = CumRuns + 1
   If CumRuns = M, Cum(5) = Cum(5) +1; go to 7
   Generate u from a uniform distribution U(0,1)
   If u<.05, go to **1**
   If u<.10, go to **2**
   If u<.20, to to **3**
   If u<.30, go to **4**
   If u<.70, go to **5**
6 CumRuns = CumRuns + 1
   If CumRuns = M, Cum(5) = Cum(6) +1; go to 7
   Generate u from a uniform distribution U(0,1)
   If u<.02 go to 5
   Go to 6
   ....
   ....
   ....
   etc.

We see how easy it is to carry out simulations of discrete-time, discrete-state processes. Add-ons of side conditions are not a problem. In the case of the viability of the king, we could just as easily deal with a matrix of transition probabilities which is variable in time. We could allow for events outside the standard model. For example, we could assume that as time progressed, domestic affairs in the United States might cut support for the king. This could be made a part of the model.

4.1.3 Screening for Cancer

The test for precancerous conditions of the cervix is quite common in the United States. Many women make it a part of their annual medical examination. Cervical cancer is associated with less than excellent sexual hygiene. Amongst observant Orthodox Jews, cervical cancer is essentially unknown.

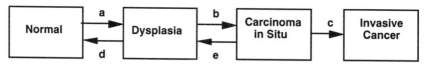

Figure 4.1. Stages to Cervical Cancer.

In the early 1970s there was significant interest in state-provided annual screenings for all women aged 20 or more. In a series of three papers, Coppleson and Brown [1–3] developed a Markov chain model which seemed to indicate that such a program would not be cost-effective. In Figure 4.1, we show their basic model.

The main goal of testing is to discover the condition in the dysplasia stage or in the cancer in situ stage. Here, a relatively nontraumatic procedure can restore the patient to the normal stage. But Coppleson and Brown noted that if the "back pressure" in the transition probabilities was relatively high, then in many cases the condition would have spontaneously corrected itself without surgery. The cost of the screenings at the time was in the $25 range. Nowadays, it would be more like $100. To confound the issue, the incidence of false negatives (that is, tests which declared the patient to be in the normal stage when she was not) was around 25%.

Let us consider below a simplification of the Coppleson–Brown transition probabilities. Let us suppose that $a = .01$, $b = .10$, $c = .05$ until age 45 and is .10 thereafter; $d = .95$, $e = .20$ until age 55 and 0 thereafter. Then, we can easily find the cost of prevention of cervical cancer by screenings annually, every five years, or according to a variable age-dependent frequency. We do this in Problem 4.2. The bottom line was that largely as a result of the Coppleson–Brown study, the decision was made to leave screenings for cervical cancer as a matter of personal choice and expense.

In this section, we note the power of simulation when dealing with discrete state models indexed on time. Getting closed form solutions is essentially hopeless whenever the transition probabilities are dependent on time and/or the history of the process. But simulation is generally a straightforward matter. Essentially, if we can describe the (time)forward process, it is generally easy to carry out many simulations, providing expected values (or histograms of states, etc.) as reasonable approximations to what we need.

4.2   Poisson Process Modeling

In 1837, well before there were the plethora of technological processes which suit his modeling strategy, Poisson [4], proposed the following model to deal with $P_k(t) = Prob[k$ events in $[0, t)]$. Everything flows from the following four axioms:

(1) $Pr[1$ event in $[t, t + h)] = \lambda h + o(h)$.
(2) $Pr[2$ or more events in $[t, t + h)] = o(h)$.
(3) $Pr[j$ events in $[t_1, s_1)$ and k in $[t_2, s_2)] =$
    $Pr[j$ in $[t_1, s_1)]Pr[k$ in $[t_2, s_2)]$ if $[t_1, s_1) \bigcap [t_2, s_2)] = \phi$.
(4) λ is constant over time.

Then

$$
\begin{aligned}
P_k(t + h) &= P_k(t)P_0(h) + P_{k-1}(t)P_1(h) \\
&= P_k(t)[1 - \lambda h + o(h)] + P_{k-1}(t)[\lambda h + o(h)],
\end{aligned}
$$

so

$$P_k(t + h) - P_k(t) = \lambda h[P_{k-1}(t) - P_k(t)] + o(h). \qquad (4.6)$$

Dividing by h and letting $h \to \infty$, we have

$$\frac{dP_k(t)}{dt} = \lambda[P_{k-1}(t) - P_k(t)]. \qquad (4.7)$$

Simple substitution in (4.7) verifies Poisson's solution:

$$P_k(t) = \frac{e^{-\lambda t}(\lambda t)^k}{k!}. \qquad (4.8)$$

The mean and variance of k are easily shown both to be equal to λt.

Let us consider an early application of Poisson's model. The German statistician von Bortkiewicz [7] examined the number of suicides of women in eight German states in 14 years. His results are shown in Table 4.1. Now, there it is an interesting question as to whether it can plausibly be claimed that the suicide data follows Poisson's model. If we compute the sample mean of the number of suicides per year, we find that it is 3.473. We can then use this value as an estimate for λt. In Table 4.1, we also show the expected numbers of suicides using the Poisson model.

Table 4.1. Actual and Expected Numbers of Suicides per Year											
Sds.	0	1	2	3	4	5	6	7	8	9	≥ 10	Sum
Freq.	9	19	17	20	15	11	8	2	3	5	3	112
E(Fq.)	3.5	12.1	21	24.3	21	14.6	8.5	4.2	1.9	.7	.2	112

One of the oldest statistical tests is Karl Pearson's *goodness of fit*. When data is naturally categorized, as it is here, in k bins (the number of suicides per state per year), if the number observed in a bin is X_i and the expected number, according to a model, is E_i, then

$$\sum_{i=1}^{k} \frac{(X_i - E_i)^2}{E_i} \approx \chi^2(k-1). \tag{4.9}$$

For the von Bortkiewicz data, we compute a value of χ^2 of 54.9. This is well beyond the limit of $\chi^2_{.990}(10)$ value of 23.21, so we might reject the applicability of the Poisson model. On the other hand, the Pearson approximation is asymptotic. We require a minimum number for each E_i of 5. In the present example, that would mean that we would have to pool the first two bins and the last four. That would give the revised Table 4.2.

Table 4.2. Actual and Expected Numbers of Suicides per Year

Suicides	≤ 1	2	3	4	5	6	≥ 7	Sum
Freq.	28	17	20	15	11	8	13	112
E(Freq.)	15.6	21	24.3	21	14.6	8.5	7	112

   This gives us a χ^2 value of 19.15, which is above the $\chi^2_{.990}(6)$ value of 16.81 but below the $\chi^2_{.998}(6)$ value of 20.79. Depending upon the use we intend to make of the Poisson model, we might choose to accept it. Yet, the relatively small sample involved might make us wish to try other approaches. For example, we know we have totals of suicides per year given in Table 4.1. We might decide to employ the following strategy:

Algorithm: Resampled Data Compared with Model-Generated Data

1. Create an "urn" with nine **0** balls, nineteen **1** balls, seventeen **2** balls, and so on.

2. With replacement, sample from the urn 1000 samples of size 112, noting the results.

3. For each of the 1000 samples, compute the χ^2 statistic in (4.9) using the original values in Table 4.1 for the E_i.

4. Using the estimate for λt of 3.473, divide the line segment from zero to 1 according to the Poisson model. Thus the probability of finding a state with zero suicides in a year is $\exp(-\lambda t) = .031$. The **0** Poisson bin then is $[0, .031)$. The probability of finding a state with one suicide in a year is $\exp(-\lambda t)\lambda t/1! = .108$. So the **1** Poisson bin is $[.031, .031 + .108)$, and so on.

5. Repeat 1000 times 112 draws of a uniform [0,1] random variable.

6. Using the E_i values from the third row in Table 4.1, compute the χ^2 statistic in (4.9).

7. Compute histograms for both the resampling simulation and that of the Poisson model. If the overlap is, say 5%, accept the hypothesis that the Poisson model fits the data.

This algorithm is an empirical preview of the resampling strategies pursued in Chapters 5 and 10.

Uses of the Poissonian framework are seen, very frequently, in the simulation of a train of time-indexed events. Now,

$$1 - F(t) = P[0 \text{ events in } [0, t)] = \exp[-\lambda t]. \tag{4.10}$$

But $F(t)$, the probability that an event occurs on or before t, is a continuous cumulative distribution function and is, by Theorem 1.1 of Chapter 1, distributed as a uniform variate on $[0, 1]$. So, also,is $1 - F(t)$. Thus, it is an easy matter, starting at time zero, to simulate the time of the next event. We generate u from $U[0, 1]$. Then the time of the next simulated event is given by

$$t = -\frac{1}{\lambda} \log(u). \tag{4.11}$$

Thus, it is possible to create a series of n simulated events by simply generating $u_1, u_2, \ldots, u_n$ and then using

$$t_i = t_{i-1} - \frac{1}{\lambda} \log(u_i). \tag{4.12}$$

Next, let us consider what might be done in equation (4.6) if we relax the axiom that states that λ must be constant. We can easily do this for the special case where we are considering $P_0(t)$, that is, the probability that no events happen in the time interval $[0, t)$

$$P_0(t + h) - P_0(t) = \lambda h[-P_0(t)] + o(h). \tag{4.13}$$

Dividing by h and taking the limit as h goes to zero, we have

$$\frac{1}{P_0(t)} \frac{dP_0(t)}{dt} = -\lambda(t). \tag{4.14}$$

Integrating from 0 to t, we have

$$P_0(t) = \exp\left[-\int_0^t \lambda(\tau)d\tau\right]. \tag{4.15}$$

We are now able to carry out simulations in rather complicated situations. Let us suppose for example [8], that a tumor, starting with one cell, grows exponentially according to

$$v(t) = ce^{\alpha t}, \text{ where } c \text{ is the volume of one cell.} \tag{4.16}$$

Next, let us suppose that this tumor will throw off metastases at a rate a proportional to the volume of the tumor. So, then the probability a metastasis will be produced on or before time t is given by

$$F_M(t) = 1 - \exp\left[-\frac{ac}{\alpha}e^{\alpha t_M}\right]. \tag{4.17}$$

From (4.17) we can easily write a simulation for the origination times of metastases starting from a tumor with given values of c, α, and a.

4.3 Solving Systems of Linear Equations

In the early days of the digital computer, von Neumann and others used Monte Carlo for the approximate solution of fairly large systems of linear equations. In this day and time where the solution of, say, 1000 equations in 1000 unknowns is doable in fairly short time on a personal computer, these techniques are not as much used as formerly. However, it continues to be the case that if the system is really quite large, say 500,000 equations in 500,000 unknowns, then Monte Carlo techniques might still be appropriate. Following largely the explications of Shreider [6] and Rubinstein [5], this section is included largely as one of historical interest. If there is one thing that has been well and truly done to death on the digital computer, it is linear algebra. As a practical matter, this is not generally an area where simulation beats classical techniques.

4.3.1 An Inefficient Procedure

Consider the system given by

$$\mathbf{AX} = \mathbf{b} \tag{4.18}$$

or

$$\sum_{k=1}^{n} a_{ik}x_k = b_i \; ; \; i = 1, 2, \ldots, n$$

where

$$\mathbf{A} = \begin{bmatrix} a_{11} & a_{12} & \cdots & a_{1n} \\ a_{21} & a_{22} & \cdots & a_{2n} \\ \vdots & \vdots & & \vdots \\ a_{n1} & a_{n2} & \cdots & a_{nn} \end{bmatrix} ; \mathbf{X} = \begin{bmatrix} x_1 \\ x_2 \\ \vdots \\ x_n \end{bmatrix} ; \mathbf{b} = \begin{bmatrix} b_1 \\ b_2 \\ \vdots \\ b_n \end{bmatrix}.$$

The solution of (4.18) may be viewed as the minimization by least squares of

$$V(\mathbf{X}) = \sum_{i=1}^{n} \alpha_i \left(\sum_{k=1}^{n} a_{ik} x_k - b_i \right)^2, \tag{4.19}$$

where the minimization holds for arbitrary $\{\alpha_j > 0\}_{j=1}^{n}$. Now, the solution $\mathbf{x}^0$ will be the center of the n-ellipsoid

$$V(\mathbf{x}) \le c. \tag{4.20}$$

We embed this ellipsoid in an n-dimensional parallelepiped given by

$$\mathbf{C} < \mathbf{X} < \mathbf{D}, \tag{4.21}$$

where

$$\mathbf{C} = \begin{bmatrix} c_1 \\ c_2 \\ \vdots \\ c_n \end{bmatrix} ; \ \mathbf{D} = \begin{bmatrix} d_1 \\ d_2 \\ \vdots \\ d_n \end{bmatrix}.$$

Let us sample $\boldsymbol{\xi}^i$ from $\mathcal{U}(\mathbf{C}, \mathbf{D})$ by selecting ξ_k^i from $\mathcal{U}(c_k, d_i)$. After sampling the n components of $\boldsymbol{\xi}^i$, we test to see whether $V(\boldsymbol{\xi}^i) \le c$. If it is, we retain the point as a sample from the n-ellipsoid, otherwise, we discard it. Thus, if $V(\boldsymbol{\xi}^i) \le c$,

$$E(\boldsymbol{\xi}^i) = \mathbf{x}^0.$$

An estimate for the accuracy of an estimator

$$\hat{\mathbf{x}}^0 = \frac{1}{N} \sum_{j=1}^{N} \boldsymbol{\xi}^j. \tag{4.22}$$

may be given by

$$\hat{\mathrm{Var}}(\hat{x}_l^0) = \frac{1}{N(N-1)} \sum_{j=1}^{N} (\xi_l^j - \hat{x}_l^0)^2. \tag{4.23}$$

This algorithm is appealing because of its generality. However, it is clearly highly inefficient, particularly when n is large.

4.3.2 An Algorithm Based on Jacobi Iteration

Next, we consider an approach based on an iterative method of Jacobi. If we wish, we may decompose the $\mathbf{A}$ matrix in (4.18) via

$$\mathbf{A} = \mathbf{E} - \mathbf{F}, \tag{4.24}$$

where $\mathbf{E}$ is invertible. Then we have as an equivalent form of (4.18)

$$\mathbf{EX} = \mathbf{FX} + \mathbf{b}. \tag{4.25}$$

We seek to implement the iterative algorithm

$$\mathbf{EX}^{(r+1)} = \mathbf{FX}^{(r)} + \mathbf{b}, \tag{4.26}$$

where $\mathbf{X}^{(r)}$ is the rth stage approximation to $\mathbf{X}^0$, and $\mathbf{X}^{(0)}$ is a first guess. Firstly, under what conditions will

$$\mathbf{X}^{(r)} - \mathbf{X} = \boldsymbol{\delta}^{(r)} \to \mathbf{0} \text{ and hence } \mathbf{X}^{(r)} \to \mathbf{X}? \tag{4.27}$$

Subtracting (4.27) from (4.26) gives

$$
\begin{aligned}
\mathbf{E}[\mathbf{X}^{(r+1)} - \mathbf{X}] &= \mathbf{F}[\mathbf{X}^{(r)} - \mathbf{X}] \\
\mathbf{E}\boldsymbol{\delta}^{(r+1)} &= \mathbf{F}\boldsymbol{\delta}^{(r)} \\
\boldsymbol{\delta}^{(r+1)} &= \mathbf{E}^{-1}\mathbf{F}\boldsymbol{\delta}^{r} \\
&= \mathbf{H}\boldsymbol{\delta}^{(r)} \text{ where } \mathbf{H} = \mathbf{E}^{-1}\mathbf{F}.
\end{aligned}
\tag{4.28}
$$

Thus

$$\boldsymbol{\delta}^{(r)} = \mathbf{H}^{r}\boldsymbol{\delta}^{(0)}. \tag{4.29}$$

Hence, if $\mathbf{H}^{r} \to \mathbf{0}$ as $r \to \infty$, then $\boldsymbol{\delta}^{(r)} \to \mathbf{0}$ and $\mathbf{X}^{(r)} \to \mathbf{X}$. Let the n linearly independent eigenvectors of $\mathbf{H}$ be denoted by $\mathbf{h}_i$, the eigenvalues by λ_i. Then we may express any $\mathbf{Y}$ as

$$\mathbf{Y} = \alpha_1\mathbf{h}_1 + \alpha_2\mathbf{h}_2 + \ldots + \alpha_n\mathbf{h}_n$$

and

$$\mathbf{H}^{r}\mathbf{Y} = \alpha_1\lambda_1^{r}\mathbf{h}_1 + \alpha_2\lambda_2^{r}\mathbf{h}_2 + \ldots + \alpha_n\lambda_n^{r}\mathbf{h}_n.$$

Hence, if $\text{Max}_i|\lambda_i| < 1$,

$$\mathbf{H}^{r}\mathbf{Y} \to \mathbf{0} \text{ as } r \to \infty.$$

But this works for all $\mathbf{Y}$, so

$$\mathbf{H}^{r} \to \mathbf{0} \text{ as } r \to \infty.$$

Thus, if the eigenvalues of $\mathbf{H} = \mathbf{E}^{-1}\mathbf{F}$ are less than unity in absolute value, the iterative scheme

$$\mathbf{X}^{(r+1)} = \mathbf{E}^{-1}\mathbf{FX}^{(r)} + \mathbf{E}^{-1}\mathbf{b} \tag{4.30}$$

converges to $\mathbf{X}$.

A common practice is to use $\mathbf{E} = \mathbf{I}$. Let us use $\mathbf{B}$ for $\mathbf{F}$ in this special case. This gives

$$\mathbf{X}^{(r+1)} = \mathbf{BX}^{(r)} + \mathbf{b} \tag{4.31}$$

and

$$\mathbf{X}^{(r)} = \mathbf{B}^{r-1}\mathbf{b} + \mathbf{B}^{(r-2)} + \ldots + \mathbf{Bb} + \mathbf{b}. \tag{4.32}$$

Then the mth coordinate of $\mathbf{X}$ is given by

$$
\begin{aligned}
x_m \;=\; & b_m + \sum_{i_1} B_{mi_1} b_{i_1} + \sum_{i_1,i_2} B_{mi_1} B_{i_1 i_2} b_{i_2} + \ldots \\
& + \sum_{i_1,i_2,\ldots,i_r} B_{mi_1} B_{i_1 i_2} \ldots B_{i_{r-1} i_r} b_{i_r} + \ldots.
\end{aligned}
\tag{4.33}
$$

We shall devise a Monte Carlo procedure to obtain an estimate for (4.33). Suppose, for a moment, that

$$\sum_{i=1}^{n} B_{mi} = 1 \text{ for all } m \text{ with all } B_{mi} \geq 0. \tag{4.34}$$

Consider a random urn sampling scheme with n distinct types of balls in each of n urns. Let the probability that a ball of the ith type is drawn from the mth urn be B_{mi}. Let us define the random variable

$\xi_m^1 = b_i$ if the ball drawn from the mth urn is of the ith type.

Clearly,

$$E(\xi_m^1) = \sum_{i_1=1}^{n} B_{mi_1} b_{i_1} \text{ , the second term in (4.33).}$$

Next, we again draw a ball from the mth urn. If the ball is of type i_1, we proceed to the i_1th urn. If we draw a type i_2 from the i_1th urn, we define the random variable

$$\xi_n^2 = b_{i_2},$$

that is, ξ_m^2 takes the value b_{i_2} with probability $\sum_{i_1=1}^{n} B_{mi_1} B_{i_1 i_2}$. Clearly, $E(\xi_m^2) = \sum_{i_1,i_2} B_{mi_1} B_{i_1 i_2} b_{i_2}$, the third term on the right hand side of (4.33). It is obvious how one proceeds to obtain successive term estimates for the right hand side of (4.33). We observe that it is necessary to relax the condition that $B_{mi} \geq 0$, $\sum_{i=1}^{n} B_{mi} = 1$, since such a condition would imply that $\mathbf{B}$ has the eigenvalue $\lambda = 1$.

Let us write the elements of $\mathbf{B}$ as follows:

$$B_{mj} = K_{mj} P_{mj} \text{ where } 0 < P_{mj} < 1 . \tag{4.35}$$

The elements of $\mathbf{b}$ are to be written

$$b_m = k_m p_m \text{ where } 0 < p_m < 1. \tag{4.36}$$

The p_m's are chosen such that

$$p_m + \sum_{j=1}^{n} P_{mj} = 1. \tag{4.37}$$

Then (4.33) becomes

$$x_m = k_m p_m + \sum_{i_1} K_{mi_1} k_{i_1} P_{mi_1} p_{i_1} + \cdots$$

$$+ \sum_{i_1 i_2 \ldots i_r} K_{mi_1} K_{i_1 i_2} K_{i_2 i_3} \ldots K_{i_{r-1} i_r} P_{mi_1} P_{i_1 i_2} \ldots P_{i_{r-1} i_r} p_{i_r} \qquad .(4.38)$$

We now assume that each of n distinct urns contains balls of $n+1$ types. The probabilities of drawing a ball of types $1, 2, \ldots, n, n+1$ from urn k are $P_{k1}, P_{k2}, \ldots, P_{kn}, p_k$, respectively. Start with urn m and draw a ball from it. If a ball of type i_1 is drawn (with $i_1 < n+1$), then the next draw is from urn i_1, and so on. However, as soon as a ball of type $n+1$ is drawn, we cease the process and give ξ_m the value $K_{mi_1} K_{i_1 i_2} K_{i_2 i_3} \ldots K_{i_{r-1}, i_r} k_{i_r}$. Now,

$$P[\xi_m] = \sum_{r=1}^{\infty} \sum_{i_1, \ldots, i_r} K_{mi_1}, \ldots, K_{i_{r-1} i_r} k_{i_r} P_{mi_1} \ldots P_{i_{r-1} i_r} p_{i_r}. \qquad (4.39)$$

Comparing with (4.38), we have

$$x_m^0 = E(\xi_m). \qquad (4.40)$$

It is of interest to note that the modified Jacobi algorithm may be used to find any component of $\mathbf{x}^0$ without computing the rest.

Problems

4.1. In the "saving the king" example,
(a) compute the probability that the king will still be in power in five years, in 10 years.
(b) Then compute the probability that the king will land in stage **6** before landing in stage **1** or stage **2**.

4.2. In the cervical cancer screening example, assuming a life expectancy of 80 years, compute the cost per case of invasive cancer prevented if screenings are carried out annually at a cost of $100, assuming a false negative rate of .25.

4.3. One of the oldest models of a stochastic process is that of Poissonian (exponential) flow. According to this model, the probability that no call will be made on a system on or before time t is given by

$$G(t) = \exp(-\lambda t),$$

where λ is a positive constant.

(a) Show that if the expected time to a call is μ, then $\lambda = 1/\mu$.

(b) From Theorem 1.1 of Chapter 1, we know that $G(t) = u$ is distributed as a uniform random variable. Using the uniform generator from (1.27), if $\lambda = 2$, simulate the first 100 calls using the algorithm

$$t_i = t_{i-1} + \frac{1}{\lambda} \log u,$$

where u is uniformly distributed on the interval $[0,1]$.

4.4. The owner of a time-sharing computer charges his customers c dollars per CPU minute. Jobs are called in according to a Poisson flow with

$$\lambda = \frac{10}{100 + c^{1.5}},$$

where λ is given in terms of calls per minute. The job times are distributed as exponential, the average time of a job being 6 minutes. If there are k jobs on the computer, the allocation utilized on each program is $1/k$ of the total. There are four phone receptacles on the computer. Discuss average profit per minute as a function of c.

4.5. In the generation of metastases given in (4.17), we will give all volume units in cells, so that $c = 1$. Time will be in months. $a = 1.7 \times 10^{-10}$, $\alpha = .31$. Using 1000 simulations, give a histogram of the times till the origination of the first metastasis.

4.6. Carry out the Algorithm: Resampled Data Compared with Model Generated Data on the von Bortkiewicz suicide data (see Section 4.2).

References

[1] Coppleson, L.W. and Brown, B.W. (1974). "Estimation of the screening error rates in repeated cervical cytology," *Amer. J. Obstet. Gynecol.*, **119**, 953.

[2] Coppleson, L.W. and Brown, B.W. (1974). " Observation on a model of the biology of carcinoma of the cervix: a poor fit between observation and theory," *Amer. J. Obstet. Gynecol.*, **122**, 127.

[3] Coppleson, L.W. and Brown, B.W. (1976). " The prevention of carcinoma of the cervix," *Amer. J. Obstet. Gynecol.*, **125**, 153.

[4] Poisson, S.D. (1837). *Recherches sur la probabilité des jugements en matière criminelle et en matière civile, précedées des réglés générales du calcul des probabilités.* Paris.

[5] Rubinstein, R.Y. (1981). *Simulation and the Monte Carlo Method.* New York: John Wiley & Sons, 158–172.

[6] Shreider, Y.A. (1966). *The Monte Carlo Method: The Method of Statistical Trials.* New York: Pergamon Press, 24–35.

[7] Stuart, A. and Ord. J.K. (1987). *Kendall's Advanced Theory of Statistics,* 5th ed., **1**. New York: Oxford University Press, 7.

[8] Thompson, J.R. and Tapia, R.A. (1990). *Nonparametric Function Estimation, Modeling, and Simulation.* Philadelphia: SIAM, 214–226.

Chapter 5

SIMEST, SIMDAT, and Pseudoreality

5.1 Computers Si, Models No

Many of us have had the experience of wishing we had 10 times the data at hand. Many of us have had the experience of trying to estimate the parameters of a model in a situation where we found it mathematically infeasible to write down a likelihood function to maximize. Many of us have needed to look at a higher-dimensional data set by trying to use our rather limited three (or four)-dimensional perceptions of reality.

I recall some years ago being on the doctoral committee of an electrical engineering student who had some time-indexed data, where the sampling intervals were so wide that he was unable to detect features at the frequencies where his interest lay. His solution (of which he was inordinately proud) was to create a spline curve-fitting algorithm which would magically transform his discrete data into continuous data. By one fell swoop he had rendered Nyquist irrelevant. Although second readers should generally keep their silence, I had to point out that he had assumed away high frequency components by his approach.

Having hosted at Rice in the late 1970s one of the early short courses on exploratory data analysis, I recall one of the two very distinguished instructors in the course making the statement that "EDA frees us from the straightjacket of models, allowing the data to speak to us unfettered by preconceived notions." During the balance of the course, whenever a new data set was displayed from one of the dual projectors, a rude psychometrician in the audience would ostentatiously cup his hand to his left ear as though to hear the data better. The psychometrician knew full well that data is perceived via a model, implicit or explicit. The strength of EDA is that it has provided us with the most effective analog–digital computer interface currently available, with the human visual system providing the analog portion of the system. Certainly, Tukey's exploratory data analysis [27] is one of the most important data analytical tools of the the last 50 years. But it has tended to tie us to a three-dimensional perception, which is unlikely to be appropriate when dealing with data of high dimensionality.

The same criticism may be made of those of us who have tried to push graphical displays of nonparametric density estimates into high-dimensional situations in the hope that spinning, coloring, and so on, might somehow force the data into a realm where our visual systems would suffice to extract the essence of the system generating the data. For many years now, I have been artistically impressed by those who would take data sets, color them, spin them, project them, time lapse them. But, in retrospect, it is hard to think of many examples where these fun type activities contributed very much to an understanding of the basic mechanism which formed the data set in the first place. Unfortunately, it seems as though many computer intensive studies, in the case of EDA, nonparametric density estimation, nonparametric regression,and so on, in the hands of many users, have more or less degenerated into essentially formalist activities, that is, activities in which we are encouraged not so much to appreciate what data analytical insights the algorithms contribute, but rather to appreciate the algorithms *sui generis*, as intrinsically wonderful.

In the matter of both density estimation and nonparametric regression, the bulk of the computer intensive work continues to emphasize the one-dimensional situation, that in which simpleminded methods (e.g., histograms, hand-fit curves, etc.) have been used for a long time with about as much success as the newer techniques. Is it not interesting to observe that the histogram is still much the most used of the one-dimensional nonparametric density estimators, and that one-dimensional curve fits are, at their most effective, psychometric activities, where one tries to automate what a human observer might do freehand? In the case where several independent variables are contemplated in nonparametric regression, rather unrealistic assumptions about additivity tend to be implemented in order to make the formal mathematics come out tractably.

In the case of one-dimensional nonparametric density estimation, a fair argument can be made that Rosenblatt obtained much of the practical knowledge we have today in his 1956 paper [18]. In the case of multivariate nonparametric density estimation, we have barely scratched the surface. Those who favor pushing three-dimensional graphical arguments into higher dimensions have made definite progress (see, e.g., Scott and Thompson [19] and Scott [20]). But others take the approach that the way to proceed is profoundly nonvisual. Boswell [4], Elliott and Thompson [10], and Thompson and Tapia [25] take the position that it is reasonable to start by seeking for centers of high density using algebraic, nongraphical, algorithms. The case where the density is heavily concentrated on curved manifolds in high-dimensional space is a natural marriage of nonparametric regression and nonparametric density estimation and has not yet received the attention one might have hoped.

The case for nonparametric regression has been eloquently and extensively advocated by Hastie and Tibshirani [15]. Other important contributions include those of Cleveland [6], Cox [7], Eubank [11], Härdle [13], and Hart and Wehrly [14]. However, I fear that the nonparametric regressor is swimming against a sea of intrinsic troubles. First of all, extrapolation continues to be both the major thing we need for application, and something very difficult to achieve, absent an investigation of the underlying model which generates the data. For interpolation purposes, it would appear that we really do not have anything more promising than locally averaged smoothers. The important case here is, of course, the one with a reasonable number of independent variables. Unfortunately, much of the work continues to be for the one-independent-variable case, where the job could be done "with a rusty nail."

In the cases mentioned above: EDA, nonparametric regression, nonparametric density estimation, algorithms have been and are being developed which seek to affect data analysis positively by use of the high-speed digital computer. I have to opine that it seems most of this "machine in search of a problem" approach has not yet been particularly successful.

5.2   The Bootstrap: A Dirac-Comb Density Estimator

There are statistical areas in which the use of the computer has borne considerable fruit. One that comes to mind is the bootstrap of Efron [8], clearly one of the most influential algorithms of the last 30 years or so. To motivate the bootstrap, we follow the discussions in [21], [22], and [25]. Let us first consider the Dirac-comb density estimator associated with a one-dimensional data set $\{x_i\}_{i=1}^n$. The Dirac-comb density estimator is given by

$$\hat{f}_\delta(x) = \frac{1}{n} \sum_{i=1}^n \delta(x - x_i). \tag{5.1}$$

We may represent $\delta(x)$ as

$$\delta(x) = \lim_{\tau \to 0} \frac{1}{\sqrt{2\pi\tau}} e^{-x^2/2\tau^2}. \tag{5.2}$$

Waving our hands briskly, $\delta(x)$ can be viewed as a density function which is zero everywhere except at the data points. At each of these points, the density has mass $1/n$. Nonparametric density estimation is frequently regarded as a subset of smoothing techniques in statistics. $\hat{f}_\delta(x)$ would appear to be infinitely rough and decidedly nonsmooth. Moreover, philosophically, it is strongly nominalist, for it says that all that can happen in any other experiment is a repeat of the data points already observed, each occurring with probability $1/n$. In other words, the data is considered to be all of reality.

For many purposes, however, $\hat{f}_\delta(x)$ is quite useful. For example, the mean of $\hat{f}_\delta(x)$ is simply

$$\mu_\delta = \int_{-\infty}^{\infty} x \hat{f}_\delta(x) = \frac{1}{n} \sum_{i=1}^{n} x_i = \bar{x}. \tag{5.3}$$

The sample variance can be represented by

$$\sigma_\delta^2 = \int_{-\infty}^{\infty} (x - \bar{x})^2 \hat{f}_\delta(x) dx = \frac{1}{n} \sum_{i=1}^{n} (x_i - \bar{x})^2 = s^2. \tag{5.4}$$

So, if we are interested in discrete characterizations of the data (such as a few lower-order moments), a Dirac comb may work quite satisfactorily. The Dirac-comb density estimator may be easily extended to higher dimensions. Indeed, such an extension is easier than is the case for other estimators, for it is based on masses at points; and a point in, say, 20-space, is still a point (hence, of zero dimension). Thus, if we have a sample of size n from a density of dimension p, $\hat{f}_\delta(x)$ becomes

$$\hat{f}_\delta(X) = \frac{1}{n} \sum_{i=1}^{n} \delta(X - X_i), \tag{5.5}$$

where

$$\delta(X) = \lim_{\tau \to 0} \left(\frac{1}{\sqrt{2\pi\tau}} \right)^p \exp\left(-\frac{\sum_{j=1}^{p} x_j^2}{2\tau^2} \right), \tag{5.6}$$

with x_j being the j component of X.

For the two-dimensional case, we might wish to develop a 95% confidence interval for the correlation coefficient,

$$\rho = \frac{\text{Cov}(x, y)}{\sigma_x \sigma_y}. \tag{5.7}$$

Now, if we had a sample of size n: $\{x_i, y_i\}_{i=1}^{n}$, we could construct

$$\hat{f}_\delta(x, y) = \frac{1}{n} \sum_{i=1}^{n} \delta((x, y) - (x_i, y_i)). \tag{5.8}$$

Next, we construct 10,000 resamplings (with replacement) of size n. That means, for each of the 10,000 resamplings we draw samples from the n data points (with replacement) of size n. For each of the resamplings, we compute the sample correlation:

$$r_j = \frac{\sum_{i=1}^{n}(x_{ji} - \overline{x}_j)(y_{ji} - \overline{y}_j)}{\sqrt{\sum_{i=1}^{n}(x_{ji} - \overline{x}_j)^2 \sum_{i=1}^{n}(y_{ji} - \overline{y}_j)^2}}. \tag{5.9}$$

Next, we can rank the sample correlations from smallest to largest. A 95% confidence interval estimate is given by

$$r_{(250)} < \rho < r_{(9,750)}. \tag{5.10}$$

One can view the bootstrap as being based on such a Dirac-comb estimator. Although it is clear that such a procedure may have use for estimating the lower moments of some interesting parameters, we should never lose sight of the fact that it is, after all, based on the profoundly discontinuous Dirac-comb estimator $\hat{f}_\delta$. The smoothed bootstrap [9] operates very much like the bootstrap itself, except that to each resampled point one adds, say, a normal variate with small variance. Essentially one samples from a fuzzy Dirac-comb nonparametric density estimator.

Now, for better or for worse, the reality is that much of statistics is concerned with such tasks as estimating a few moments. When we know the underlying density function (and history shows that people get away with assuming that the world is Gaussian more often than might be supposed), then knowledge of a few moments actually gives a continuous description of the underlying system [the first and second moments of a Gaussian (normal) distribution completely characterize the density function everywhere].

However, if the world truly were Gaussian, then we could drop the entire subject of nonparametrics (and most of the computer-intensive statistical analyses). Let us consider an example where the data really is Gaussian, but the use of the Dirac-comb nonparametric density estimator serves us ill. For example, suppose we have a sample of size 100 of firings at a bull's-eye of radius 5 centimeters. If the distribution of the shots is circular normal with mean the center of the bull's-eye and deviation 1 meter, then with a probability in excess of .88, none of the shots will hit inside the bull's-eye. Then any Dirac-comb resampling procedure will tell us the bull's-eye is a safe place if we get a base sample (as we probably will) with no shots in the bullseye. Such a problem with the bootstrap motivated SIMDAT, the nonparametric density estimator based resampling algorithm of Taylor and Thompson [21, 22, 25].

5.3 SIMDAT: A Smooth Resampling Algorithm

We note that any realization of a bootstrap simulation most likely will be different from the original sample. Some of the sample points will disappear. Others will be repeated multiple times. Indeed, the concatenation of a bootstrap followed by a bootstrap based on that bootstrapped simulation, and so on, will lead ultimately to a simulated sample which consists of a single sample point. This is hardly desirable. It might be hoped that a single resampling would be of such a character that we would be almost indifferent as to whether we had this simulation or the original data set. But, of course, it would be dangerous to wander too far from the original sample. A resampling of a resampling of a resampling, and so on, is not nearly as desirable as resamples which always point directly to the original sample.

The bootstrap is clearly a powerful algorithm for many purposes. However, given the ubiquity of fast computing, it would usually be preferred to use resampling schemes based on better nonparametric density estimators than the Dirac comb. One such would be the 1976 algorithm of Guerra, Tapia and Thompson [12], where one obtains a smooth of the empirical cdf and samples from that. This algorithm has been employed for some time as the RNGCT subroutine of Visual Numerics (formerly IMSL). The disadvantage of the algorithm is that it was only written for the one-dimensional case, and that the estimator of the cdf must be explicitly obtained.

One candidate for a nonparametric density estimator to be used for simulation purposes would be

$$\hat{f}_\delta(X) = \frac{1}{n} \sum_{i=1}^{n} K(X - X_i, \Sigma_i), \tag{5.11}$$

where $K(X, \Sigma_i)$ is a normal distribution centered at zero with locally estimated covariance matrix Σ_i.

Such an estimator, despite its advantages, would appear to be very difficult to construct. However, let us recall what it is we seek: not a nonparametric density estimator, but a random sample from such an estimator. So, perhaps, we can go directly from the actual sample to the pseudosample. Of course, this is precisely what the bootstrap estimator does, with the frequently unfortunate properties associated with a Dirac comb. Fortunately, it is possible to go from the sample directly to the pseudosample in such a way that the resulting estimator behaves very much like that of the normal kernel approach above. This is what the SIMDAT algorithm does.

5.3.1   The SIMDAT Algorithm

Assume given a data set of size n from a p-dimensional variable X, $\{X_i\}_{i=1}^n$. Assume that we have already rescaled our data set so that the marginal sample variances in each vector component are the same. For a given integer m, we can find, for each of the n data points, the $m - 1$ nearest neighbors. These will be stored in an array of size $n \times (m - 1)$.

Suppose we wish to generate a pseudosample of size N. Note that there is no reason to suppose that n and N need be the same (as is the case generally with the bootstrap). To start the algorithm, we sample one of the n data points with probability $1/n$ (just as with the bootstrap). Then, we recall its $m-1$ nearest neighbors from memory, and compute the mean of the resulting set of m points:

$$\overline{X} = \frac{1}{m} \sum_{i=1}^{m} X_i. \tag{5.12}$$

Next, we subtract from each of the data points the local mean $\overline{X}$, thus achieving zero averages of the transformed cloud:

$$\{X'_j\} = \{X_j - \overline{X}\}_{j=1}^{m}. \tag{5.13}$$

Although we go through the computations of sample means and centering about them here as though they were a part of the simulation process, the operation will be done once only, just as with the determination of the $m-1$ nearest neighbors of each data point. The $\{X'_j\}$ values as well as the $\overline{X}$ values will be stored in an array of dimension $n \times (m+1)$.

Next, we generate a random sample of size m from the one-dimensional uniform distribution:

$$U\left(\frac{1}{m} - \sqrt{\frac{3(m-1)}{m^2}}, \frac{1}{m} + \sqrt{\frac{3(m-1)}{m^2}}\right). \tag{5.14}$$

We now generate our centered pseudodata point X', via

$$X' = \sum_{l=1}^{m} u_l X'_l. \tag{5.15}$$

Finally, we add back on $\overline{X}$ to obtain our pseudodata point X:

$$X = X' + \overline{X}. \tag{5.16}$$

These, then, are the nuts and bolts of SIMDAT. The major setup cost is the determination of interpoint distances. The tablulation is for each of the n data points, a list indicating the $m-1$ nearest points. Once the resulting matrix has been obtained, subsequent generation of any desired amount of pseudodata is very rapid.

5.3.2   An Empirical Justification of SIMDAT

As m and n get large, the procedure gives results very much like those of the normal kernel approach mentioned earlier. To see why this is so, we consider the sampled vector X_l and its $m-1$ nearest neighbors:

$$\{X_l\}_{l=1}^m = \begin{bmatrix} x_{1l} \\ x_{2l} \\ \vdots \\ x_{kl} \end{bmatrix}_{l=1,\dots,m}. \tag{5.17}$$

Let us treat this collection of m points as being from a distribution with mean vector μ and covariance matrix Σ. Now, if $\{u_l\}_{l=1}^m$ is an independent sample from the uniform distribution in (5.14), then

$$E(u_l) = \frac{1}{m}; \ \text{Var}(u_l) = \frac{m-1}{m^2}; \ \text{Cov}(u_i, u_j) = 0, \ \text{for } i \neq j. \tag{5.18}$$

Then we form the linear combination

$$Z = \sum_{l=1}^m u_l X_l. \tag{5.19}$$

We note that for the rth component of the vector Z, $z_r = u_1 x_{r1} + u_2 x_{r2} + \dots + u_m x_{rm}$,

$$E(z_r) = \mu_r, \tag{5.20}$$

$$\text{Var}(z_r) = \sigma_r^2 + \frac{m-1}{m}\mu_r^2, \tag{5.21}$$

and

$$\text{Cov}(z_r, z_s) = \sigma_{rs} + \frac{m-1}{m}\mu_r \mu_s. \tag{5.22}$$

We observe that if the mean vector of X were $(0, 0, \dots, 0)$, then the mean vector and covariance matrix of Z would be the same as that of X, that is, $E(z_r) = 0$, $\text{Var}(z_r) = \sigma_r^2$, and $\text{Cov}(z_r, z_s) = \sigma_{rs}$. Naturally, by translation to the local sample mean of the nearest-neighbor cloud, we will not quite have achieved this result. But we will come very close to the generation of an observation from the truncated distribution that generated the points in the nearest-neighbor cloud.

For m moderately large, by the central limit theorem, SIMDAT comes close to sampling from n normal distributions with the mean and covariance matrices corresponding to those of the n, m nearest-neighbor clouds. If we were seeking rules for consistency of the nonparametric density estimator corresponding to SIMDAT, we could use the formula of Mack and Rosenblatt [16] for nearest-neighbor nonparametric density estimators:

$$m = Cn^{4/(p+4)}. \tag{5.23}$$

Actually, as a practical matter, such formulas have little practical relevance, since C is usually not available. Furthermore, we ought to remember that our goal is not to obtain a nonparametric density estimator, but rather, to generate a data set which appears like that of the data set before us. Let us suppose that we err on the side of making m far too small, namely, $m = 1$. That would yield simply the bootstrap. Suppose that we err on the side of making m far too large, namely, $m = n$. That would yield an estimator which roughly sampled from a multivariate normal distribution with the mean vector and covariance matrix computed from the data.

In Figure 5.1 we show a sample of size 85 from a mixture of three normal distributions with the weights indicated, and a pseudodata set of size 85 generated by SIMDAT with $m = 5$. We note that the emulation of the data is reasonably good. In Figure 5.2 we go through the same exercise, but with $m = 15$. There, effects of a modest oversmoothing are noted. In general, if the data set is very large, say of size 1000 or greater, good results are obtained with $m \approx .02n$. For smaller values of n, m values in the $.05n$ range appear to work well. A version of SIMDAT in the S language, written by E.N. Atkinson, is available under the name "gendat" from **http://lib.stat.cmu.edu/S/gendat**. (For really large data sets, the user may wish to use Fortran or C instead of S. The savings for using the more primitive languages, as opposed to S, may be twentyfold. The coding of SIMDAT can easily be done in less than 100 lines of Fortran or C code.)

$$\mu_1 = \begin{bmatrix} -1 \\ -2 \end{bmatrix}; \ \mu_2 = \begin{bmatrix} -2 \\ 3 \end{bmatrix}; \ \mu_3 = \begin{bmatrix} 2 \\ 3 \\ 2 \end{bmatrix}. \qquad \Sigma_1 = \begin{bmatrix} 1 & \frac{1}{2} \\ -\frac{1}{2} & 1 \end{bmatrix}; \ \Sigma_2 = \begin{bmatrix} 1 & \frac{1}{2} \\ \frac{1}{2} & 1 \end{bmatrix}; \ \Sigma_3 = \begin{bmatrix} 1 & \frac{1}{10} \\ \frac{1}{10} & 1 \end{bmatrix}.$$

$$\text{Mixed Distribution} = \frac{1}{2}N_1 + \frac{1}{3}N_2 + \frac{1}{6}N_3.$$

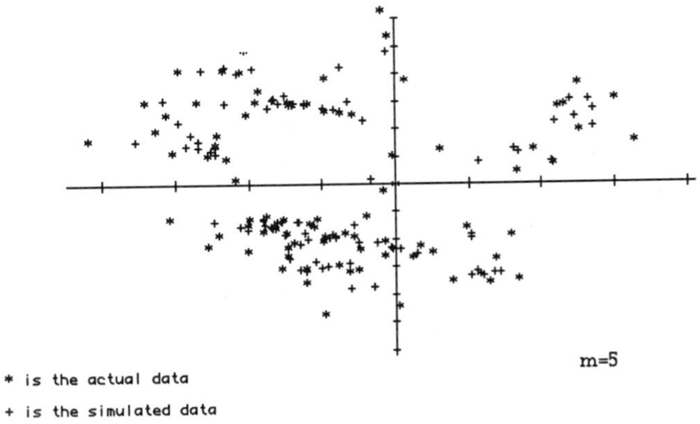

m=5

* is the actual data

\+ is the simulated data

Figure 5.1. Undersmoothed SIMDAT.

$$\mu_1 = \begin{bmatrix} -1 \\ -2 \end{bmatrix}; \ \mu_2 = \begin{bmatrix} -2 \\ 3 \end{bmatrix}; \ \mu_3 = \begin{bmatrix} 2 \\ \frac{3}{2} \end{bmatrix}. \qquad \Sigma_1 = \begin{bmatrix} 1 & \frac{1}{2} \\ -\frac{1}{2} & 1 \end{bmatrix}; \ \Sigma_2 = \begin{bmatrix} 1 & \frac{1}{2} \\ \frac{1}{2} & 1 \end{bmatrix}; \ \Sigma_3 = \begin{bmatrix} 1 & \frac{1}{10} \\ \frac{1}{10} & 1 \end{bmatrix}.$$

$$\text{Mixed Distribution} = \frac{1}{2} N_1 + \frac{1}{3} N_2 + \frac{1}{6} N_3.$$

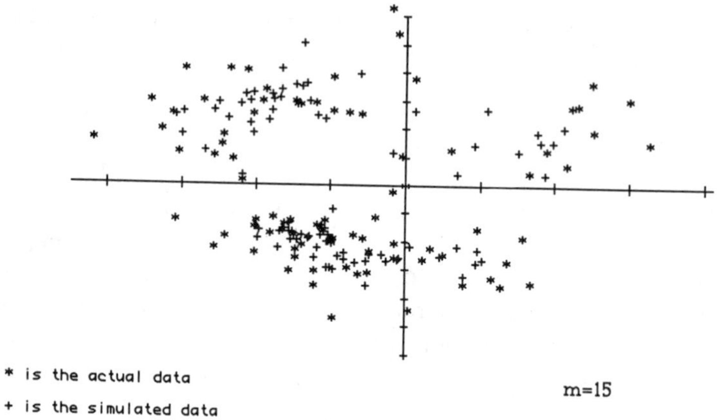

* is the actual data

+ is the simulated data

m=15

Figure 5.2. Oversmoothed SIMDAT.

So far, we have considered basically model-free techniques for examining data. There are, of course, many situations where exploration of a new data set may preclude an early conjecture as to a likely model of the mechanism generating the data. In my opinion, such procedures should usually be first steps in modeling a process. But, unfortunately, they frequently are as far as one goes. In essence, the nonparametric techniques use the power of the computer to bypass altogether the need for the modeling step. Such an approach is likely to be useful mainly as an interpolative device. When the dimensionality of a data set becomes high, say five or more, this *adhocery* is likely to prove dangerous, since we may be confronted with a number of widely separated modes, with deserts in between. Dealing with such data sets, nonparametrically, away from the modes, is an extrapolation problem, and using the standard smoothed interpolation routines can bring one quickly to disaster.

5.4 SIMEST: An Oncological Example

The power of the computer as an aid to modeling does not get the attention it deserves. Part of the reason is that the human modeling approach tends to be analog rather than digital. Analog computers were replaced by digital computers 40 years ago. Most statisticians remain fascinated by the graphical capabilities of the digital computer. The exploratory data analysis route tends to attempt to replace modeling by visual displays which are then interpreted, in a more-or-less instinctive fashion, by an observer. Statisticians who proceed in this way are functioning somewhat like prototypical cyborgs. After over two decades of seeing data spun, colored, and graphed in a myriad of ways, I have to admit to being disappointed when comparing the promise of EDA with its reality. Its influence amongst academic statisticians has been enormous. Visualization is clearly one of the major areas in the statistical literature. But the inferences drawn from these visualizations in the real world are, relatively speaking, not so numerous. Moreover, when visualization-based inferences are drawn, they tend to give results one might have obtained by classical techniques.

Of course, as in the case of using the computer as a nonparametric smoother, some uses are better than others. Around a decade ago a group of Bayesian statisticians convinced one of our leading research universities that the reason statistical analysis had produced marginal results in such areas as oncology had been the traditional dominance of frequentists in biometry. The advent of high-speed computing brought forth the possibility that the insights of physicians could be appropriately blended into priors leading to breakthrough posteriors. Here, we were told that the computer would enable us to carry out another one, two, or three dimensions of quadrature, thus enabling prior information to be infused into the process. But, since the desired prior information really was not available (and may never be in the form required), the computer just enabled people to spin their wheels faster.

It is extremely unfortunate that some are so multicultural in their outlook that they rearrange their research agenda in order to accommodate themselves to our analog-challenged friends, the digital computers. Perhaps the greatest disappointment is to see the modeling aspect of our analog friends, the human beings, being disregarded in favor of using them as gestaltic image processors. This really will not do. We need to rearrange the agenda so that the human beings can gain the maximal assistance from the computers in making inferences from data. That is the purpose of SIMEST.

There is an old adage to the effect that quantitative change carried far enough may produce qualitative change. The fact is that we now have computers so fast and cheap that we can proceed (almost) as though computation were free and instantaneous (with infinite accessible memory thrown in as well). This should change, fundamentally, the way we approach data analysis in the light of models.

There are now a number of examples in several fields where SIMEST has been used to obtain estimates of the parameters characterizing a market-related applied stochastic process (see, e.g., Bridges, Ensor, and Thompson [5]). Below we consider an oncological application to motivate and to explicate SIMEST. We shall first show a traditional model-based data analysis, note the serious (generally insurmountable) difficulties involved, and then give a simulation-based, highly computer-intensive way to get what we require to understand the process and act upon that understanding.

5.4.1   An Exploratory Prelude

In the late 1970s, my colleague Barry W. Brown, of the University of Texas M.D. Anderson Cancer Center, and I had started to investigate some conjectures concerning reasons for the relatively poor performance of oncology in the American "War on Cancer." Huge amounts of resources had been spent with less encouraging results than one might have hoped. It was my view that part of the reason might be that the basic orthodoxy for cancer progression was, somehow, flawed.

This basic orthodoxy can be summarized briefly as follows

> At some time, for some reason, a single cell goes wild. It, and its progeny, multiply at rates greater than that required for replacement. The tumor thus formed grows more or less exponentially. From time to time, a cell may break off (metastasize) from the tumor and start up a new tumor at some distance from the primary (original) tumor. The objective of treatment is to find and excise the primary before it has had a chance to form metastases. If this is done, then the surgeon (or radiologist) will have "gotten it all" and the patient is cured. If metastases are formed before the primary is removed, then a cure is unlikely, but the life of the patient may be extended and ameliorated by aggressive administration of chemotherapeutic agents which will kill tumor cells more vigorously than normal cells. Unfortunately, since the agents do attack normal cells as well, a cure of metastasized cancer is unlikely, since the patient's body cannot sustain the dosage required to kill all the cancer cells.

For some cancers, breast cancer, for example, long-term cure rates had not improved very much for many years.

5.4.2   Model and Algorithms

One conjecture, consistent with a roughly constant intensity of display of secondary tumors, is that a patient with a tumor of a particular type is not displaying breakaway colonies only, but also new primary tumors due to suppression of a patient's immune system to attack tumors of a particular type. We can formulate axioms at the micro level which will incorporate the mechanism of new primaries. Such an axiomitization has been formulated by Bartoszyński, Brown, and Thompson [3]. The first five axioms are consistent with the classical view as to metastatic progression. Hypothesis 6 is the mechanism we introduce to explain the nonincreasing intensity function of secondary tumor display.

Hypothesis 1. For any patient, each tumor originates from a single cell and grows at exponential rate α.

Hypothesis 2. The probability that the primary tumor will be detected and removed in $[t, t + \Delta t)$ is given by $bY_0(t)\Delta t + o(\Delta t)$, and until the removal of the primary, the probability of a metastasis in $[t, t + \Delta t)$ is $aY_0(t)\Delta t + o(\Delta t)$, where $Y_0(t)$ is the size of the primary tumor at time t.

Hypothesis 3. For patients with no discovery of secondary tumors in the time of observation, S, put $m_1(t) = Y_1(t) + Y_2(t) + \ldots$, where $Y_i(t)$ is the size of the ith originating tumor. After removal of the primary, the probability of a metastasis in $[t, t + \Delta t)$ equals $am_1(t) + o(\Delta t)$, and the probability of detection of a new tumor in $[t, t + \Delta t)$, is $bm_1(t) + o(\Delta t)$.

Hypothesis 4. For patients who do display a secondary tumor, after removal of the primary and before removal of Y_1, the probability of detection of a tumor in $[t, t + \Delta t)$ equals $bY_1(t) + o(\Delta t)$, while the probability of detection of a metastasis is $aY_1(t) + o(\Delta t)$.

Hypothesis 5. For patients who do display a secondary tumor, the probability of a metastasis in $[t, t + \Delta t)$ is $am_2(t)\Delta t + o(\Delta t)$, while the probability of detection of a tumor is $bm_2(t)\Delta t + o(\Delta t)$, where $m_2(t) = Y_2(t) + \ldots$.

Hypothesis 6. The probability of a systemic occurrence of a tumor in $[t, t + \Delta t)$ equals $\lambda \Delta t + o(\Delta t)$, independent of the prior history of the patient.

Essentially, we shall attempt to develop the likelihood function for this model so that we can find the values of $a, b, \alpha,$ and λ which maximize the likelihood of the data set observed. It turns out that this is a formidable task indeed. The SIMEST algorithm which we develop later gives a quick alternative to finding the likelihood function. However, to give the reader some feel as to the complexity associated with model aggregation from seemingly innocent axioms, we shall give some of the details of getting the likelihood function. First of all, it turns out that in order to have any hope of obtaining a reasonable approximation to the likelihood function, we will have to make some further simplifying assumptions. We shall refer to the period prior to detection of the primary as Phase 0. Phase 1 is the period from detection of the primary to S', the first time of detection of a secondary tumor. For those patients without a secondary tumor, Phase 1 is the time of observation, S. Phase 2 is the time, if any, between S' and S. Now for the two simplifying axioms. T_0 is defined to be the (unobservable) time between the origination of the primary and the time when it is detected and removed (at time $t = 0$). T_1 and T_2 are the times until detection and removal of the first and second of the subsequent tumors (times to be counted from $t = 0$). We shall let X be the total mass of all tumors other than the primary at $t = 0$.

Hypothesis 7. For patients who do not display a secondary tumor, growth of the primary tumor, and of all tumors in Phase 1, is deterministically exponential with the growth of all other tumors treated as a pure birth process.

Hypothesis 8. For patients who display a secondary tumor, the growth of the following tumors is treated as deterministic: in Phase 0, tumors $Y_0(t)$ and $Y_1(t)$; in Phase 1, tumor $Y_1(t)$ and all tumors which originated in Phase 0; in Phase 2, all tumors. The growth of remaining tumors in Phases 0 and 1 is treated as a pure birth process.

We now define

$$H(s; t, z) = \exp\{\frac{az}{\alpha} e^{\alpha t}(e^s - 1)\log[1 + (e^{-\alpha t} - 1)e^{-s}]$$
$$+ \frac{\lambda}{\alpha}s - \frac{\lambda}{\alpha}\log[1 + e^{\alpha t}(e^s - 1)]\} \qquad (5.24)$$

and

$$p(t; z) = bze^{\alpha t}\exp\left[-\frac{bz}{\alpha}(e^{\alpha t} - 1)\right]. \qquad (5.25)$$

Further, we shall define

$$w(y) = \lambda\left[\int_0^y e^{-\nu(u)}du - y\right], \qquad (5.26)$$

where $\nu(u)$ is determined from

$$u = \int_0^\nu (a + b + \alpha s - ae^{-s})^{-1}ds. \qquad (5.27)$$

Then, we can establish the following propositions, and from these, the likelihood function:

$$p(T_0 > \tau) = \exp\left[-b\int_0^\tau e^{\alpha t}dt\right] = \exp\left[-\frac{b}{\alpha}(e^{\alpha \tau} - 1)\right]. \qquad (5.28)$$

For patients who do not display a secondary tumor, we have

$$P(T_1 > S | X = x) = \exp\left[-x\nu(S) + w(S)\right]. \qquad (5.29)$$

For patients who develop metastases, we have

$$P(T_1 > S) = P(\text{no secondary tumor in } (0, S))$$
$$= \int_0^\infty e^{w(s)}p(t; 1)H(\nu(s); t, 1)dt. \qquad (5.30)$$

Similarly, for patients who do display a secondary tumor, we have

$$P(T_1 = S', T_2 > S) = \int_0^\infty \int_0^t e^{w(S-S')} p(t;1) p(S'; e^{\alpha u})(\lambda + a e^{\alpha(t-u)})$$
$$\times \exp\left[-\lambda(t-u) - \frac{a}{\alpha}(e^{\alpha(t-u)} - 1)\right] H(\nu(S-S'); S', e^{\alpha u})$$
$$\times H(\nu(S-S') e^{\alpha S'}; u, e^{\alpha(t-u)}) du dt$$
$$+ \int_0^\infty \int_0^{S'} e^{w(S-S')} p(t;1) \exp\left[-\lambda t - \frac{a}{\alpha}(e^{\alpha t} - 1)\right] \lambda e^{-\lambda u}$$
$$\times p(S' - u; 1) H(\nu(S-S'); S' - u, 1) du dt \quad (5.31)$$

Finding the likelihood function, even a quadrature approximation to it, is more than difficult. Furthermore, current symbol manipulation programs (e.g., Mathematica, Maple) do not have the capability of doing the work. Accordingly, it must be done by hand. Approximately 1.5 person years were required to obtain a quadrature approximation to the likelihood. Before starting this activity, we had no idea of the rather practical difficulties involved. However, the activity was not without reward.

We found estimates for the parameter values using a data set consisting of 116 women who presented with primary breast cancer at the Curie-Sklodowska Cancer Institute in Warsaw (time units in months, volume units in cells): $a = .17 \times 10^{-9}$, $b = .23 \times 10^{-8}$, $\alpha = .31$, and $\lambda = .0030$. Using these parameter values, we found excellent agreement between the proportion free of metastasis versus time obtained from the data and that obtained from the model, using the parameter values given above. When we tried to fit the model to the data with the constraint that $\lambda = 0$ (that is, disregarding the systemic process as is generally done in oncology), the attempt failed.

One thing one always expects from a model-based approach is that, once the relevant parameters have been estimated, many things one had not planned to look for can be found. For example, tumor doubling time is 2.2 months. The median time from primary origination to detection is 59.2 months and at this time the tumor consists of 9.3×10^7 cells. The probability of metastasis prior to detection of the primary is .069, and so on. A model-based approach generally yields such serendipitous results, as a nonparametric approach generally does not. It is worth mentioning that, more frequently than one realizes, we need an analysis which is flexible, in the event that at some future time we need to answer questions different from those originally posed. The quadrature approximation of the likelihood is relatively inflexible compared to the simulation-based approach we shall develop shortly.

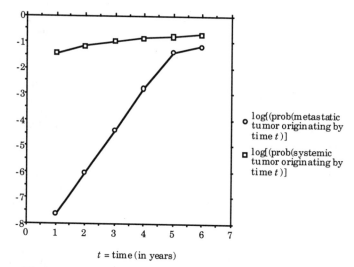

t = time (in years)

Figure 5.3. Metastatic and Systemic Effects.

Insofar as the relative importance of the systemic and metastatic mechanisms, in causing secondary tumors associated with breast cancer, it would appear from Figure 5.3 that the systemic is the more important. This result is surprising, but is consistent with what we have seen in our exploratory analysis of another tumor system (melanoma). Interestingly, it is by no means true that for all tumor systems the systemic term has such dominance. For primary lung cancer, for example, the metastatic term appears to be far more important.

It is not clear how to postulate, in any definitive fashion, a procedure for testing the null hypothesis of the existence of a systemic mechanism in the progression of cancer. We have already noted that when we suppress the systemic hypothesis, we cannot obtain even a poor maximum likelihood fit to the data. However, someone might argue that a different set of nonsystemic axioms should have been proposed. Obviously, we cannot state that it is simply impossible to manage a good fit without the systemic hypothesis. However, it is true that the nonsystemic axioms we have proposed are a fair statement of traditional suppositions as to the growth and spread of cancer.

As a practical matter, we had to use data that were oriented toward the life of the patient rather than toward the life of a tumor system. This is due to the fact that human *in vivo* cancer data is seldom collected with an idea toward modeling tumor systems. For a number of reasons, including the difficulty mentioned in obtaining the likelihood function, deep stochastic modeling has not traditionally been employed by many investigators in oncology. Modeling frequently precedes the collection of the kinds of data of greatest use in the estimation of the parameters of the model. Anyone who has gone through a modeling exercise such as that covered in this section is very likely to treat such an exercise as a once in a lifetime experience. It simply is too frustrating to have to go through all the flailing around to come up with a quadrature approximation to the likelihood function. As soon as a supposed likelihood function has been found, and a corresponding parameter estimation algorithm constructed, the investigator begins a rather lengthy "debugging" experience. The algorithm's failure to work might be due to any number of reasons (e.g., a poor approximation to the likelihood function, a poor quadrature routine, a mistake in the code of the algorithm, inappropriateness of the model,etc). Typically, the debugging process is time consuming and difficult. If one is to have any hope for coming up with a successful model-based investigation, an alternative to the likelihood procedure for aggregation must be found.

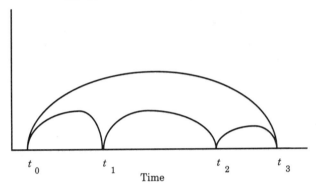

Figure 5.4. Two Possible Paths from Primary to Secondary.

In order to decide how best to construct an algorithm for parameter estimation which does not have the difficulties associated with the classical closed-form approach, we should try to see just what causes the difficulty with the classical method of aggregating from the microaxioms to the macro level, where the data lives. A glance at Figure 5.4 reveals the problem with the closed-form approach.

The axioms of tumor growth and spread are easy enough to implement in the forward direction. Indeed, they follow the natural forward formulation used since Poisson's work of 1837 [17]. Essentially, we are overlaying stochastic processes, one on top of the other, and interdependently to boot. But when we go through the task of finding the likelihood, we are essentially seeking all possible paths by which the observables could have been generated. The secondary tumor, originating at time t_3, could have been thrown off from the primary at time t_3, or it could have been thrown off from a tumor which itself was thrown off from another tumor at time t_2 which itself was thrown off from a tumor at time t_1 from the primary which originated at time t_0. The number of possibilities is, of course, infinite.

In other words, the problem with the classical likelihood approach in the present context is that it is a backward look from a database generated in the forward direction. To scientists before the present generation of fast, cheap computers, the backward approach was, essentially, unavoidable unless one avoided such problems (a popular way out of the dilemma). However, we need not be so restricted.

Once we realize the difficulty when one uses a backward approach with a concatenation of forwardly axiomitized mechanisms, the way out of our difficulty is rather clear [1, 23]. We need to analyze the data using a forward formulation. The most obvious way to carry this out is to pick a guess for the underlying vector of parameters, put this guess in the micro-axiomitized model and simulate many times of appearance of secondary tumors. Then, we can compare the set of simulated quasidata with that of the actual data.

The greater the concordance, the better we will believe we have done in our guess for the underlying parameters. If we can quantitize this measure of concordance, then we will have a means for guiding us in our next guess. One such way to carry this out would be to order the secondary occurrences in the data set from smallest to largest and divide them into k bins, each with the same proportion of the data. Then, we could note the proportions of quasidata points in each of the bins. If the proportions observed for the quasidata, corresponding to parameter value Θ, were denoted by $\{\pi_j(\Theta)\}_{j=1}^k$, then a Pearson goodness-of-fit statistic would be given by

$$\chi^2(\Theta) = \sum_{j=1}^{k} \frac{(\pi_j(\Theta) - 1/k)^2}{\pi_j(\Theta)}. \tag{5.32}$$

The minimization of $\chi^2(\Theta)$ provides us with a means of estimating Θ.

Typically, the sample size, n, of the data will be much less than N, the size of the simulated quasidata. With mild regularity conditions, assuming there is only one local maximum of the likelihood function, Θ_0, as $n \to \infty$ (which function we of course do not know), then as $N \to \infty$, as n becomes large and k increases in such a way that $\lim_{n\to\infty} k = \infty$ and $\lim_{n\to\infty} k/n = 0$, the minimum χ^2 estimator for Θ_0 will have an expected mean square error which approaches the expected mean square error of the maximum likelihood estimator. This is, obviously, quite a bonus. Essentially, we will be able to forfeit the possibility of knowing the likelihood function and still obtain an estimator with asymptotic efficiency equal to that of the maximum likelihood estimator. The price to be paid is the acquisition of a computer swift enough and cheap enough to carry out a very great number, N, of simulations, say 10,000. This ability to use the computer to get us out of the "backward trap" is a potent but, as yet seldom used, bonus of the computer age. Currently, the author is using SIMEST on a 400 MHz personal computer, amply adequate for the task, which now costs around $1000.

First, we observe how the forward approach enables us to eliminate those hypotheses which were, essentially a practical necessity if a likelihood function was to be obtained. Our new axioms are simply:

Hypothesis 1. For any patient, each tumor originates from a single cell and grows at exponential rate α.

Hypothesis 2. The probability that the primary tumor will be detected and removed in $[t, t+\Delta t)$ is given by $bY_0(t)\Delta t + o(\Delta t)$. The probability that a tumor of size $Y(t)$ will be detected in $[t, t+\Delta t)$ is given by $bY(t)\Delta t + o(\Delta t)$.

Hypothesis 3. The probability of a metastasis in $[t, t + \Delta)$ is $a\Delta t \times$ (total tumor mass present).

Hypothesis 4. The probability of a systemic occurrence of a tumor in $[t, t + \Delta t)$ equals $\lambda \Delta t + o(\Delta t)$, independent of the prior history of the patient.

In order to simulate, for a given value of (α, a, b, λ), a quasidata set of secondary tumors, we must first define:

t_D = time of detection of primary tumor;

t_M = time of origin of first metastasis;

t_S = time of origin of first systemic tumor;

t_R = time of origin of first recurrent tumor;

t_d = time from t_R to detection of first recurrent tumor;

t_{DR} = time from t_D to detection of first recurrent tumor.

Now, generating a random number u from the uniform distribution on the unit interval, if $F(\cdot)$ is the appropriate cumulative distribution function for a time, t, we set $t = F^{-1}(u)$. Then, assuming that the tumor volume at time t is

$$v(t) = ce^{\alpha t}, \text{ where } c \text{ is the volume of one cell,} \tag{5.33}$$

we have

$$F_M(t) = 1 - \exp\left(-\frac{ac}{\alpha}e^{\alpha t_M}\right). \tag{5.34}$$

Similarly, we have

$$F_D(t_D) = 1 - \exp\left(-\int_0^{t_D} bce^{\alpha \tau}\, d\tau\right)$$

$$= 1 - \exp\left(-\frac{bc}{\alpha}e^{\alpha t_D}\right), \tag{5.35}$$

$$F_S = 1 - e^{-\lambda t_S}, \tag{5.36}$$

and

$$F_d(t_d) = 1 - \exp\left(-\frac{bc}{\alpha}e^{\alpha t_d}\right). \tag{5.37}$$

Using the actual times of discovery of secondary tumors $t_1 \leq t_2 \leq, \ldots, \leq t_n$ we generate k bins. In actual tumor situations, because of recording protocols, we may not be able to put the same number of secondary tumors in each bin. Let us suppose that the observed proportions are given by $(p_1, p_2, \ldots, p_k)$. We shall generate N recurrences $s_1 < s_2 < \ldots < s_N$. The observed proportions in each of the bins will be denoted $\pi_1, \pi_2, \ldots, \pi_k$. The goodness of fit corresponding to (α, λ, a, b) will be given by

$$\chi^2(\alpha, \lambda, a, b) = \sum_{j=1}^{k} \frac{(\pi_j(\alpha, \lambda, a, b) - p_j)^2}{\pi_j(\alpha, \lambda, a, b)}. \qquad (5.38)$$

As a practical matter, we may replace $\pi_j(\alpha, \lambda, a, b)$ by p_j, since with (α, λ, a, b) far away from truth, $\pi_j(\alpha, \lambda, a, b)$ may well be zero. Then the following algorithm generates the times of detection of quasisecondary tumors for the particular parameter value (α, λ, a, b).

Secondary Tumor Simulation (α, λ, a, b)

Generate t_D
$j = 0$
$i = 0$
Repeat until $t_M(j) > t_D$
$j = j + 1$
Generate $t_M(j)$
Generate $t_{dM}(j)$
$t_{dM}(j) \leftarrow t_{dM}(j) + t_M(j)$
If $t_{dM}(j) < t_D$, then $t_{dM}(j) \leftarrow \infty$
Repeat until $t_S > 10t_D$
$i = i + 1$
Generate $t_{dS}(i)$
$t_{dS}(i) \leftarrow t_{dS}(i) + t_S(i)$
$s \leftarrow \min [t_{dM}(j), t_{dS}(i)]$
Return s
End Repeat

The algorithm above does still have some simplifying assumptions. For example, we assume that metastases of metastases will probably not be detected before the metastases themselves. We assume that the primary will be detected before a metastasis, and so on. Note, however, that the algorithm utilizes much less restrictive simplifying assumptions than those which led to the terms of the closed-form likelihood such as (5.31). Even more importantly, the Secondary Tumor Simulation algorithm can be discerned in a few minutes, whereas a likelihood argument is frequently the work of months.

Another advantage of the forward simulation approach is its ease of modification. Those who are familiar with "backward" approaches based on the likelihood or the moment generating function are only too familiar with the experience of a slight modification causing the investigator to go back to the start and begin anew. This is again a consequence of the tangles required to be examined if a backward approach is used. However, a modification of the axioms generally causes slight inconvenience to the forward simulator.

For example, we might add:

Hypothesis 5. A fraction γ of the patients ceases to be at systemic risk at the time of removal of the primary tumor if no secondary tumors exist at that time. A fraction $1 - \gamma$ of the patients remain at systemic risk throughout their lives.

Clearly, adding Hypothesis 5 will cause considerable work if we insist on using the classical aggregation approach of maximum likelihood. However, in the forward simulation method we simply add the following lines to the Secondary Tumor Simulation code:

> Generate u from $U(0, 1)$
> If $u > \gamma$, then proceed as in the Secondary Tumor Simulation code
> If $u < \gamma$, then proceed as in the Secondary Tumor Simulation code except replace the step "Repeat until $t_S > 10t_D$" with the step "Repeat until $t_S(i) > t_D$."

In the discussion of metastasis and systemic occurrence of secondary tumors, we have used a model supported by data to try to gain some insight into a part of the complexities of the progression of cancer in a patient. Perhaps this sort of approach should be termed *speculative data analysis*. In the current example, we were guided by a nonparametric intensity function estimate [2], which was surprisingly nonincreasing, to conjecture a model, which enabled us to test systemic origin against metastatic origin on something like a level playing field. The fit without the systemic term was so bad that anything like a comparison of goodness-of-fit statistics was unnecessary.

It is interesting to note that the implementation of SIMEST is generally faster on the computer than working through the estimation with the closed-form likelihood. In the four-parameter oncological example we have considered here, the running time of SIMEST was 10% of the likelihood approach. As a very practical matter, then, the simulation-based approach would appear to majorize that of the closed-form likelihood method in virtually all particulars. The running time for SIMEST can begin to become a problem as the dimensionality of the response variable increases past one. Up to this point, we have been working with the situation where the data consists of failure times. In the systemic versus metastatic oncogenesis example, we managed to estimate four parameters based on this kind of one-dimensional data. As a practical matter, for tumor data, the estimation of five or six parameters for failure time data is the most one can hope for. Indeed, in the oncogenesis example, we begin to observe the beginnings of singularity for four parameters, due to a near trade-off between the parameters a and b. Clearly, it is to our advantage to be able to increase the dimensionality of our observables. For example, with cancer data, it would be to our advantage to utilize not only the time from primary diagnosis and removal to secondary discovery and removal, but also the tumor volumes of the primary and the secondary. Such information enables one to postulate more individual growth rates for each patient. Thus, it is now appropriate to address the question of dealing with multivariate response data.

Gaussian Template Criterion. In many cases, it will be possible to employ a procedure using a criterion function. Such a procedure has proved quite successful in another context (see [26], pp. 275–280). First, we transform the data $\{X_i\}_{i=1}^n$ by a linear transformation such that for the transformed data set $\{U_i\}_{i=1}^n$ the mean vector becomes zero and the covariance matrix becomes I:

$$U = AX + b. \tag{5.39}$$

Then, for the current best guess for Θ, we simulate a quasidata set of size N. Next, we apply the same transformation to the quasidata set $\{Y_j(\Theta)\}_{j=1}^N$, yielding $\{Z_j(\Theta)\}_{j=1}^N$. Assuming that both the actual data set and the simulated data set come from the same density, the likelihood ratio $\Lambda(\Theta)$ should increase as Θ gets closer to the value of Θ, say Θ_0, which gave rise to the actual data, where,

$$\Lambda(\Theta) = \frac{\prod_{i=1}^n \exp[-\frac{1}{2}(u_{1i}^2 + \ldots + u_{pi}^2)]}{\prod_{i=1}^N \exp[-\frac{1}{2}(z_{1i}^2 + \ldots + z_{pi}^2)]}. \tag{5.40}$$

As soon as we have a criterion function, we are able to develop an algorithm for estimating Θ_0. The closer Θ is to Θ_0, the smaller will $\Lambda(\Theta)$ tend to be.

The procedure above which uses a single Gaussian template will work well in many cases where the data has one distinguishable center and a falling off away from that center which is not too "taily." However, there will be cases where we cannot quite get away with such a simple approach. For example, it is possible that a data set may have several distinguishable modes and/or exhibit very heavy tails. In such a case, we may be well advised to try a more local approach. Suppose that we pick one of the n data points at random—say x_1—and find the m nearest-neighbors amongst the data. We then treat this m nearest-neighbor cloud as if it came from a Gaussian distribution centered at the sample mean of the cloud and with covariance matrix estimated from the cloud. We transform these $m+1$ points to zero mean and identity covariance matrix, via

$$U = A_1 X + b_1. \tag{5.41}$$

Now, from our simulated set of N points, we find the $N(m+1)/n$ simulated points nearest to the mean of the $m+1$ actual data points. This will give us an expression like

$$\Lambda_1(\Theta) = \frac{\prod_{i=1}^{m+1} \exp[-\frac{1}{2}(u_{1i}^2 + \ldots + u_{pi}^2)]}{\prod_{i=1}^{N(m+1)/n} \exp[-\frac{1}{2}(z_{1i}^2 + \ldots + z_{pi}^2)]}. \tag{5.42}$$

If we repeat this operation for each of the n data points, we will have a set of local likelihood ratios $\{\Lambda_1, \Lambda_2, \ldots, \Lambda_n\}$. Then one natural measure of concordance of the simulated data with the actual data would be

$$\Lambda(\Theta) = \sum_{i=1}^{n} \log(\Lambda_i(\Theta)). \tag{5.43}$$

We note that this procedure is not equivalent to one based on density estimation, since the nearest-neighbor ellipsoids are not disjoint. Nevertheless, we have a level playing field for each of the guesses for Θ and the resulting simulated data sets.

A Simple Counting Criterion. Fast computing notwithstanding, with n in the 1000 range and N around 10,000, the template procedure can become prohibitively time consuming. Accordingly, we may opt for a subset counting procedure:

> For data size n, pick a smaller value, say nn.
>
> Pick a random subset of the data points of size nn.
>
> Pick a nearest neighbor outreach parameter m, typically $.02n$.
>
> For each of the nn data points, X_j, find the Euclidean distance to the mth nearest neighbor, say $d_{j,m}$.
>
> For an assumed value of the vector parameter Θ, generate N simulated observations.
>
> For each of the data points in the random subset of the data, find the number of simulated observations within $d_{j,m}$, say $N_{j,m}$.
>
> Then the criterion function becomes

$$\chi^2(\Theta) = \sum_{j=1}^{nn} \frac{((m+1)/n - N_{j,m}/N)^2}{(m+1)/n}.$$

Experience indicates that whatever nn size subset of the data points is selected should be retained throughout the changes of Θ. Otherwise, practical instability may obscure the path to the minimum value of the criterion function.

A SIMDAT-SIMEST Stopping Rule. In Section 4.2 we considered a situation where we compared the results from resampled data points with those from model-based simulations. SIMDAT is not a simple resampling so much as it is a stochastic interpolator. We can take the original data and use SIMDAT to generate a SIMDAT pseudodata set of N values.

Then, for a particular guess of Θ, we can compute a SIMEST pseudodata set of N values. For any region of the space of the vector observable, the number of SIMEST-generated points should be approximately equal to the number of SIMDAT-generated points. For example, let us suppose that we pick nn of the n original data points and find the radius $d_{j,m}$ of the hypersphere which includes m of the data points for, say, point X_j. Let $N_{j,SD}$ be the number of SIMDAT-generated points falling inside the hypersphere and $N_{j,SE}$ be the number of SIMEST-generated points falling inside the hypersphere. Consider the empirical goodness-of-fit statistic for the SIMDAT cloud about point X_j:

$$\chi^2_{j,SD}(\Theta) = \frac{((m+1)/n - N_{j,SD}/N)^2}{(m+1)/n}.$$

For the SIMEST cloud, we have

$$\chi^2_{j,SE}(\Theta) = \frac{((m+1)/n - N_{j,SE}/N)^2}{(m+1)/n}.$$

If the model is correct and if our estimate for Θ is correct, then $\chi^2_{j,SE}(\Theta)$ should be, on the average, distributed similarly to $\chi^2_{j,SE}(\Theta)$. Accordingly, we can construct a sign test. To do so, let

$$\begin{aligned} W_j &= \quad +1 \text{ if } \chi^2_{j,SD}(\Theta) \geq \chi^2_{j,SE}(\Theta) \\ &= \quad -1 \text{ if } \chi^2_{j,SD}(\Theta) < \chi^2_{j,SE}(\Theta). \end{aligned}$$

So, if we let

$$Z = \frac{\sum_{j=1}^{nn} W_j}{\sqrt{nn}},$$

we might decide to terminate our search for estimating Θ when the absolute value of Z falls below 3 or 4.

Problems

5.1. Generate a sample of size 100 of firings at a bull's-eye of radius 5 centimeters where the distribution of the shots is circular normal with mean the center of the bullseye and deviation 1 meter.

(a) Generate and display a bootstrapped sample of size 1000. Do you find any simulated points inside the bull's-eye?

(b) Then using $m= 10$, generate a SIMDAT pseudosample of size 1000. Do you find any simulated points inside the bull's-eye?

5.2. A multivariate distribution with heavy tales may be generated as follows. First, we generate a χ^2 variable v with 2 degrees of freedom.

Then we generate p independent univariate normal variates from a normal distribution with mean 0 and variance 1. Then $\mathbf{X}' = (X_1, X_2, \ldots, X_p)$ will have the multivariate normal distribution $\mathcal{N}_p(\mathbf{0}, \mathbf{I})$. Moreover,

$$t_{2,p}(\mu) = \frac{\mathbf{X}}{\sqrt{v/2}} + \mu$$

will have a shifted $\mathbf{t}$ distribution with 2 degrees of freedom. Generate a sample of size 500 from the mixture distribution

$$f = .9\mathcal{N}_3(\mathbf{0}, \mathbf{I}) + .1t_{2,3}(\mathbf{0}).$$

Can you tell the difference between the sample from the mixture distribution above and a sample of size 500 from $\mathcal{N}_3(\mathbf{0}, \mathbf{I})$?

5.3. Let us return to von Bortkiewicz's suicide tabulation from Chapter 4, namely Table 4.1.

Table 4.1. Actual and Expected Numbers of Suicides per Year											
Suicides	0	1	2	3	4	5	6	7	8	9	≥ 10	Sum
Freq.	9	19	17	20	15	11	8	2	3	5	3	112
E(freq.)	3.5	12.1	21	24.3	21	14.6	8.5	4.2	1.9	.7	.2	112

Using the model for k, the number of suicides generated in a year,

$$P(k|\theta) = e^{-\theta}\frac{\theta^k}{k!},$$

find a SIMEST estimator for θ, using as the criterion function

$$\sum_{i=0}^{10} \frac{(X_i - E_i)^2}{E_i} \approx \chi^2(k-1).$$

(We recall here that the category "10" here is ≥ 10.)

5.4. Generally speaking, before using an algorithm for parameter estimation on a set of data, it is best to use it on a set of data simulated from the proposed model. Returning to the problem in Section 5.4.1, generate a set of times of discovery of secondary tumor (time measured in months past discovery and removal of primary) of 400 patients with $a = .17 \times 10^{-9}$, $b = .23 \times 10^{-8}$, $\alpha = .31$, and $\lambda = .0030$. Using SIMEST, see if you can recover the true parameter values from various starting values.

5.5. Using the parameter values given in Problem 5.4, generate a set of times of discovery of first observed secondary tumor and second observed secondary tumor. Using SIMEST, see whether you can recover the true parameter values from various starting values.

References

[1] Atkinson, E.N., Bartoszyński, R., Brown, B.W., and Thompson, J.R. (1983). "Simulation techniques for parameter estimation in tumor related stochastic processes," *Proceedings of the 1983 Computer Simulation Conference.* New York: North-Holland, 754–757.

[2] Bartoszyński, Robert, Brown, B.W., McBride, C.M., and Thompson, J.R. (1981). "Some nonparametric techniques for estimating the intensity function of a cancer related nonstationary Poisson process," *Ann. Statist.,* **9**, 1050–1060.

[3] Bartoszyński, R., Brown, B.W. and Thompson, J.R. (1982). "Metastatic and systemic factors in neoplastic progression," in *Probability Models and Cancer*, L. LeCam and J. Neyman, eds. New York: North-Holland., 253–264.

[4] Boswell, S.B. (1983). *Nonparametric Mode Estimators for Higher Dimensional Densities.* Doctoral Dissertation. Houston: Rice University.

[5] Bridges, E., Ensor, K.B., and Thompson, J.R. (1992). "Marketplace competition in the personal computer industry," *Decis. Sci.,* **23**, 467–477.

[6] Cleveland, W.S. (1979). "Robust locally-weighted regression and smoothing scatterplots," *J. Amer. Statist. Assoc.,* **74**, 829–836.

[7] Cox, D.D. (1983). "Asymptotics for M-type smoothing splines," *Ann. Statist.,* **17**, 530–551.

[8] Efron, B. (1979). "Bootstrap methods–another look at the jacknife," *Ann. Statist.,* **7**, 1–26.

[9] Efron, B. and Tibshirani, R.J. (1993). *An Introduction to the Bootstrap.* New York: Chapman & Hall.

[10] Elliott, M.N. and Thompson, J.R. (1993). "An exploratory algorithm for the estimation of mode location and numerosity in multidimensional data," *Proceedings of the Thirty-Eighth Conference on the Design of Experiments in Army Research Development and Testing*, B. Bodt, ed. Reseach Triangle Park, N.C.: Army Research Office, 229–244.

[11] Eubank, R.L. (1988). *Smoothing Splines and Nonparametric Regression.* New York: Marcel Dekker.

[12] Guerra, V., Tapia, R.A., and Thompson, J. R. (1976). "A random number generator for continuous random variables based on interpolation procedure of Akima's," *Comput. Sci. Statist.,* 228–230.

[13] Härdle, W. (1990). *Applied Nonparametric Regression*. New York: Oxford University Press.

[14] Hart, J.D. and Wehrly, T.E. (1986) "Kernel regression estimation using repeated measurements data," *J. Amer. Statist. Asssoc.*, **81**, 1080−1088.

[15] Hastie, T.J. and Tibshirani, R.J.(1990). *Generalized Additive Models*. New York: Chapman & Hall.

[16] Mack, Y.P. and Rosenblatt, Murray (1979). "Multivariate k-nearest neighbor density estimates," *J. Multivariate Analysis*, **9**, 1−15.

[17] Poisson, S.D. (1837). *Recherches sur la probabilité des jugements en matière criminelle et en matière civile, précedées des réglés générales du calcul des probabilités*. Paris.

[18] Rosenblatt, M. (1956). "Remarks on some nonparametric estimates of a density function," *Ann. Math. Statist.*, **27**, 832−835.

[19] Scott, D.W. and Thompson, J.R. (1983). "Probability density estimation in higher dimensions," in *Computer Science and Statistics*, J. Gentle, ed., Amsterdam: North Holland, 173−179.

[20] Scott, D.W. (1992). *Multivariate Density Estimation*. New York: John Wiley & Sons.

[21] Taylor, M.S. and Thompson, J.R. (1982). "A data based random number generator for a multivariate distribution," *Proceedings of the NASA Workshop on Density Estimation and Function Smoothing*. College Station, Texas: Texas A & M University Department of Statistics, 214−225.

[22] Taylor, M.S. and Thompson, J.R. (1986). "A data based algorithm for the generation of random vectors," *Comp. Statist. Data Anal*, **4**, 93−101.

[23] Thompson, J.R., Atkinson, E.N., and Brown, B.W. (1987). "SIMEST: an algorithm for simulation based estimation of parameters characterizing a stochastic process," in *Cancer Modeling*. Thompson, J. and Brown, B., eds. New York: Marcel Dekker. 387−415.

[24] Thompson, J.R. (1989). *Empirical Model Building*. New York: John Wiley & Sons, 35−43.

[25] Thompson, J.R. and Tapia, R.A. (1990). *Nonparametric Function Estimation, Modeling, and Simulation*. Philadelphia: SIAM, 154−161, 214−233.

[26] Thompson, J.R. and Koronacki, J. (1993).*Statistical Process Control for Quality Improvement.* New York: Chapman & Hall.

[27] Tukey, J.W. (1977). *Exploratory Data Analysis.* Reading, Mass.: Addison-Wesley.

Chapter 6

Models for Stocks and Derivatives

6.1   Introduction

Over the years the schemes developed for "beating the stock market" have rivaled in number those for winning at gambling casinos. There were always to be seen stock managers who could boast of 20% growth in their funds over the time of existence of the funds. Over a period of years, however, it was noted that for every superperforming stock guru, there were many others who had performed well for a time, but then had run into a period of bad performance for their funds. It was the winners who received publicity rather than the losers, just as it is the person who hits the jackpot in a casino about whom we hear. It is held by some that the market is so efficient that, at any given time, the valuation of a stock takes account of the likely future performance of the stock. Hence, one could argue that one stock is pretty much as good a buy as another. Then, there is the fact that mutual funds are generally uninterested in small capitalization stocks, whereas individual investors may well be. Thus, there must be a different mechanism for buying cap stocks and large cap stocks. Then, there is the beauty contest argument originally posed by Keynes and more recently advanced by Soros that an intelligent buyer of stocks is always trying to access not which stocks he or she thinks are the best, but which stocks other investors will consider the best. By this argument, Coca-Cola Enterprises, whose price/earnings ratio soars up to 100, is considered to be a good investment, because Coca-Cola will not be a stock other people will want to dump. And so on. The arguments become quite convoluted. One is reminded of the old Sherlock Holmes scenario where Professor Moriarty has made his escape on the last train to Dover. But then Holmes can hire a special train to pursue him. But then Moriarty knows that Holmes will hire a special train, so he departs from the train before Dover. But then Holmes knows that Moriarty knows that he (Holmes) can hire a special train and that, consequently, it is a good idea to presume that Moriarty will not catch the packet from Dover to Calais. But then Moriarty knows .... Although some (e.g., George Soros) have done amazingly well using a variation of the psychoinvesting strategy, the road is strewn with the silent corpses of those who were not successful at it.

Major private universities have, at times, neared insolvency by putting their endowments into the hands of this or that guru. For example, 30 years ago, some bought onto the notion that the total stock and bond assets of an endowment could be expended just as though they were interest payments. The university could safely expend, say, 5% of everything that accrued from its portfolio, including the stocks themselves. The growth of the market would enable continued growth in the endowment without particular worries. And since the assets would be invested broadly across the market, there would really be no significant chance that a bad patch of stock performances would bring the endowment to grief.

For some time now, many have argued that most grandiose models do not work very well. It was embarrassing to brokerage houses when, starting around 20 years ago, some persons actually started selecting portfolios by throwing darts at a board containing the market listings. The embarrassment was accentuated when these portfolios frequently outperformed those of the experts. If one assumes perfect flow of information and rationality on the part of investors, it can indeed be argued that all news, good and bad, for a particular stock is instantaneously reflected in the price of the stock. That being so, it could be argued that the movement of the market is essentially a random walk. Of course, information is neither perfect nor instantaneously assimilated, and investors are not completely rational, but the random walk argument has a great deal of intuitive appeal based on past history.

Generally, one assumes that the value of a random selection of stocks will grow with time. Clearly, the assumption is based, in large measure, on technological advances, broadly interpreted. What is the "fair rate of return on investment"? Answers range from the classical Marxist position of zero to complicated formulas based on levels of risk.

Let us consider the difference equation

$$\frac{\Delta X}{X} = \mu \Delta t. \tag{6.1}$$

Letting the time spacing go to zero gives us simply the law of continuous compound interest with growth rate μ:

$$X(t) = X(0)e^{\mu t}. \tag{6.2}$$

Alternatively, let us consider

$$\frac{\Delta X}{X} = \mu \Delta t + \sigma \epsilon \sqrt{\Delta t}. \tag{6.3}$$

Here σ (the *volatility*) is a measure of the variability of the process as time increases. ϵ is a normal variate with mean zero and variance 1. In the limit, as Δt goes to zero, such a process is uniquely defined and is commonly referred to as a geometric Brownian process.

$$d(\ln X) = \mu dt + \sigma dz. \tag{6.4}$$

6.2   Ito's Lemma

Let us suppose we have a continuously differentiable function of two variables $G(x,t)$. Then, taking a Taylor's expansion through terms of the second order, we have

$$\Delta G \quad \approx \quad \frac{\partial G}{\partial x} \Delta x + \frac{\partial G}{\partial t} \Delta t$$
$$+ \frac{1}{2} \frac{\partial^2 G}{\partial x^2} (\Delta x)^2 + \frac{1}{2} \frac{\partial^2 G}{\partial t^2} (\Delta t)^2 + \frac{\partial^2 G}{\partial x \partial t} \Delta x \Delta t. \tag{6.5}$$

Next let us consider the *general Ito process*

$$dx = a(x,t)dt + b(x,t)dz \tag{6.6}$$

with discrete version

$$\Delta x = a(x,t)\Delta t + b(x,t)\epsilon\sqrt{\Delta t} \tag{6.7}$$

where dz denotes a Wiener process, and a and b are deterministic functions of x and t. We note that

$$(\Delta x)^2 = b^2 \epsilon^2 \Delta t + \text{ terms of higher order in } \Delta t. \tag{6.8}$$

Now
$$\text{Var}(\epsilon) = E(\epsilon^2) - [E(\epsilon)]^2 = 1.$$
So, since by assumption $E(\epsilon) = 0$,
$$E(\epsilon^2) = 1.$$

Furthermore, since ϵ is $\mathcal{N}(0,1)$, after a little algebra, we have that $\text{Var}(\epsilon^2) = 2$, and $\text{Var}(\Delta t \epsilon^2) = 2(\Delta t)^2$. Thus, if Δt is very small, through terms of order $(\Delta t)^2$, we have that it is equal to its expected value, namely,

$$(\Delta x)^2 = b^2 \Delta t. \tag{6.9}$$

Substituting (6.7) and (6.9) into (6.5), we have *Ito's Lemma*

$$\Delta G = \left(\frac{\partial G}{\partial x} a(x,t) + \frac{\partial G}{\partial t} + \frac{1}{2} \frac{\partial^2 G}{\partial x^2} b^2 \right) \Delta t + \frac{\partial G}{\partial x} b\epsilon\sqrt{\Delta t} \tag{6.10}$$

or

$$dG = \left(\frac{\partial G}{\partial x} a(x,t) + \frac{\partial G}{\partial t} + \frac{1}{2} \frac{\partial^2 G}{\partial x^2} b^2 \right) dt + \frac{\partial G}{\partial x} b dz. \tag{6.11}$$

6.3   A Geometric Brownian Model for Stocks

Following Hull [3] and Smith [5], let us consider the situation where a stock price follows geometric Brownian motion:

$$dS = \mu S dt + \sigma S dz. \tag{6.12}$$

Now, in Ito's lemma we define $G = \ln S$. Then we have

$$\frac{\partial G}{\partial S} = \frac{1}{S}; \; \frac{\partial^2 G}{\partial S^2} = -\frac{1}{S^2} \; ; \; \frac{\partial G}{\partial t} = 0.$$

Thus G follows a Wiener process:

$$dG = \left(\mu - \frac{\sigma^2}{2} \right) dt + \sigma dz. \tag{6.13}$$

This tells us simply that if the price of the stock at present is given by $S(0)$, then the value t units in the future will be given by

$$
\begin{aligned}
S(t) \; &= \; S(0) \exp \left[(\mu - \frac{\sigma^2}{2}) t + \epsilon \sigma \sqrt{t} \right] \\
&= \; S(0) \exp \left[\mathcal{N} \left((\mu - \frac{\sigma^2}{2}) t, t\sigma^2 \right) \right] \\
&= \; \exp \left[\mathcal{N} \left(\log(S(0)) + (\mu - \frac{\sigma^2}{2}) t, t\sigma^2 \right) \right],
\end{aligned}
\tag{6.14}
$$

where $\mathcal{N}(\log(S(0)) + (\mu - \sigma^2/2)t, t\sigma^2)$ is a normal random variable with mean $\log(S(0)) + (\mu - \sigma^2/2)t$ and variance $t\sigma^2$. Thus, $S(t)$ is a normal variable exponentiated (i.e., it follows the *lognormal* distribution). The expectation of $S(t)$ is given by $S(0) \exp[\mu t]$. In the current context, the assumption of an underlying geometric Brownian process (and hence that $S(t)$ follow a lognormal distribution) is somewhat natural. Let us suppose we consider the prices of a stock at times t_1, $t_1 + t_2$, and $t_1 + t_2 + t_3$. Then if we assume $S(t_1 + t_2)/S(t_1)$ to be independent of starting time t_1, and if we assume $S(t_1 + t_2)/S(t_1)$ to be independent of $S(t_1 + t_2 + t_3)/S(t_1 + t_2)$, and if we assume the variance of the stock price is finite for finite time, and if we assume that the price of the stock cannot drop to zero, then, it can be shown that $S(t)$ must follow geometric Brownian motion and have the lognormal distribution indicated.

From (6.14), we have, for all t and Δt

$$r(t + \Delta t, t) = \frac{S(t + \Delta t)}{S(t)} = \exp \left[\left(\mu - \frac{\sigma^2}{2} \right) \Delta t + \epsilon \sigma \sqrt{\Delta t} \right]. \tag{6.15}$$

Defining $R(t + \Delta t, t) = \log(r(t + \Delta t, t))$, we have

$$R(t + \Delta t, t) = \left(\mu - \frac{\sigma^2}{2} \right) \Delta t + \epsilon \sigma \sqrt{\Delta t}.$$

Then

$$E[R(t + \Delta t, t)] = \left(\mu - \frac{\sigma^2}{2}\right)\Delta t. \tag{6.16}$$

Suppose we have a stock that stands at 100 at week zero. In 26 subsequent weeks we note the performance of the stock as shown in Table 6.1. Here $\Delta t = 1/52$. Let

$$\bar{R} = \frac{1}{26}\sum_{i=1}^{26} R(i) = .1524.$$

Table 6.1. 26 Weeks of Stock Performance

Week=i	Stock(i)	$r(i)$=Stock(i)/Stock($i-1$)	$R(i)$=log($r(i)$)
1	99.83942	0.99839	
2	97.66142	0.97818	-0.02206
3	97.54407	0.99880	-0.00120
4	96.24717	0.98670	-0.01338
5	98.65675	1.02503	0.02473
6	102.30830	1.03701	0.03634
7	103.82212	1.01480	0.01469
8	103.91875	1.00093	0.00093
9	105.11467	1.01151	0.01144
10	104.95000	0.99843	-0.00157
11	105.56152	1.00583	0.00581
12	105.44247	0.99887	-0.00113
13	104.21446	0.98835	-0.01171
14	103.58197	0.99393	-0.00609
15	102.70383	0.99152	-0.00851
16	102.94174	1.00232	0.00231
17	105.32943	1.02320	0.02293
18	105.90627	1.00548	0.00546
19	103.63793	0.97858	-0.02165
20	102.96025	0.99346	-0.00656
21	103.39027	1.00418	0.00417
22	107.18351	1.03669	0.03603
23	106.02782	0.98922	-0.01084
24	106.63995	1.00577	0.00576
25	105.13506	0.98589	-0.01421
26	107.92604	1.02655	0.02620

By the strong law of large numbers, the sample mean $\bar{R}$ converges almost surely to its expectation $(\mu - \sigma^2/2)\Delta t$. Next, we note that

$$[R(t + \Delta t, t) - E(R(t + \Delta t, t))]^2 = \epsilon^2\sigma^2\Delta t, \tag{6.17}$$

so

$$\text{Var}[R(t + \Delta t, t)] = E[R(t + \Delta t, t) - \left(\mu - \frac{\sigma^2}{2}\right)\Delta t]^2 = \sigma^2\Delta t. \tag{6.18}$$

For a large number of weeks, this variance is closely approximated by the sample variance

$$s_R^2 = \frac{1}{26-1} \sum_{i=1}^{26} (R(i) - \bar{R})^2 = .000258.$$

Then $\hat{\sigma}^2 = .000258/\Delta t = .000258 \times 52 = .013416$, giving as our volatility estimate $\hat{\sigma} = .1158$. Finally, our estimate for the growth rate is given by

$$\hat{\mu} = \bar{R} + \frac{\hat{\sigma}^2}{2} = .1524 + .0067 = .1591.$$

6.4   Diversification

We will now consider strategies for reducing the risk of an investor.

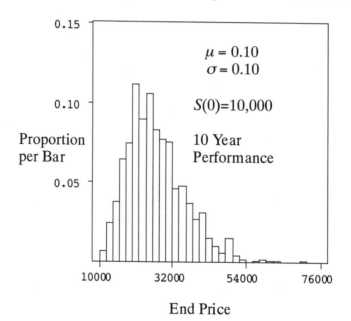

Figure 6.1. Histogram of Common Stock End Prices.

Let us consider a histogram of 1000 possible outcomes of an investment of $10,000 in a stock with $\mu = .10$ and $\sigma = .10$ as shown in Figure 6.1. If we can find 20 stocks, each with $\mu = .10$ and $\sigma = .10$, then *assuming they are stochastically independent of each other*, we might take the $10,000 and invest $500 in each of the stocks. The histogram of performance (using 200 possible outcomes) is shown in Figure 6.2. The sample means for both the one-stock investment and the diversified 20-stock mutual fund are

27,088 and 26,846, respectively. But the standard deviation for the one-stock investment (8,731) is roughly $\sqrt{20}$ times that for the mutual fund investment (1875). A portfolio that has such such stochastic independence would be a truly *diversified* one. Generally speaking, one should expect some dependency between the tracks of the stocks.

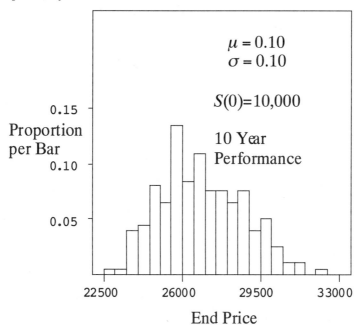

Figure 6.2. Histogram of Idealized Mutual Fund End Prices.

Let us modify (6.3) to allow for a mechanism for dependence:

$$\frac{\Delta S_i}{S_i} = \mu \Delta t + \sigma \epsilon_i \sqrt{\Delta t}. \tag{6.19}$$

We shall take η_0 to be a Gaussian random variable with mean zero and variance 1. Similarly, the 20 η_i will also be independent Gaussian with mean zero and variance 1. Then we shall let

$$\epsilon_i = c(a\eta_0 + (1-a)\eta_i). \tag{6.20}$$

We wish to select c and a so that a is between zero and 1 and so that $\text{Var}(\epsilon_i) = 1$ and any two ϵ_i and ϵ_j have positive correlation ρ. After a little algebra, we see that this is achieved when

$$a = \frac{\rho - \sqrt{\rho(1-\rho)}}{2\rho - 1} \tag{6.21}$$

and

$$c^2 = \frac{1}{a^2 + (1-a)^2}. \tag{6.22}$$

At the singular value of $\rho = .5$, we use $a = .5$. Let us examine the situation with an initial stake of \$500 per stock with $\mu = \sigma = .10$ and $\rho = .8$ as shown in Figure 6.3. We employ 500 simulations. We note that the standard deviation of the portfolio has grown to 7,747. This roughly follows the rule that the standard deviation of a portfolio where stocks have the same variance and have correlation ρ, should be $\sqrt{1 + (n-1)\rho}$ times that of an uncorrelated portfolio.

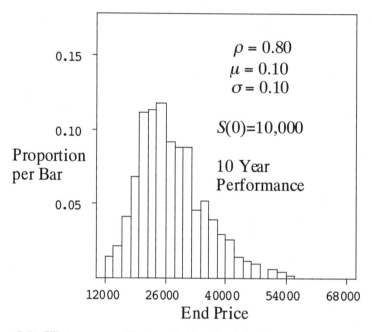

Figure 6.3. Histogram of Mutual Fund with Correlated Stock Prices.

6.5 Negatively Correlated Portfolios

Is there anything more likely to reduce the variance of a mutual fund portfolio than the assumption that the stocks move in a stochastically independent fashion? We recall that if we have two random variables X_1 and X_2, each with unit variance and the same unknown mean μ, the variance of the sample mean is given by

$$\text{Var}((X_1 + X_2)/2) = \frac{1}{4}[2 + 2\rho]. \qquad (6.23)$$

Here the variance can be reduced to zero if $\rho = -1$. Let us consider a situation where we have two stocks each of value \$5000 at time zero which grow according to

$$\frac{\Delta S_1}{S_1} = \mu\Delta t + \sigma\epsilon\sqrt{\Delta t} \qquad (6.24)$$

$$\frac{\Delta S_2}{S_2} = \mu\Delta t - \sigma\epsilon\sqrt{\Delta t},$$

where ϵ is a Gaussian variate with mean zero and unit variance. Then the resulting portfolio (based on 500 simulations) is exhibited in Figure 6.4.

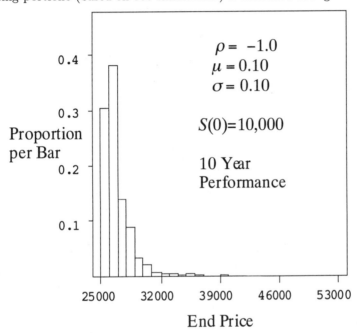

Figure 6.4. **Histogram of Two-Stock Portfolio with $\rho = -1$.**

We note that the standard deviation of this two-stock portfolio is 1701, even less than that observed for the 20-stock portfolio with the assumption of independence of stocks. Now, the assumption that we can actually find stocks with negative correlation to the tune of -1 is unrealistic. Probably, we can find two stocks with rather large negative correlation, however. This is easily simulated via

$$\frac{\Delta S_1}{S_1} = \mu\Delta t + (a\epsilon_0 + (1 - a)\epsilon_1)c\sqrt{\Delta t}\sigma \qquad (6.25)$$

$$\frac{\Delta S_i}{S_2} = \mu\Delta t - (a\epsilon_0 + (1 - a)\epsilon_2)c\sqrt{\Delta t}\sigma,$$

where

$$a = \frac{\rho + \sqrt{-\rho(1 - \rho)}}{2\rho + 1},$$

$c = 1/\sqrt{a^2 + (1-a)^2}$ and ϵ_0, ϵ_1, and ϵ_2 are normally and independently distributed with mean zero and variance 1. Let us consider, in Figure 6.5, the situation where $\rho = -.5$.

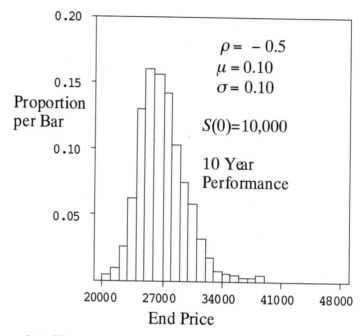

Figure 6.5. Histogram of Two-Stock Portfolio with $\rho = -.5$.

We note that the standard deviation here has grown to 2719. When it comes to utilizing negative correlation as a device for the reduction of the variance of a portfolio, a number of strategies can be considered. We know, for example, that if one wishes to minimize the variance of a sample mean, we can pose the problem as a constrained optimization problem to find the optimal correlation matrix, where we impose the constraint that the covariance matrix be positive definite. Our problem here is rather different, of course.

We could try something simple, namely take our two-stock negatively correlated portfolio and repeat it 10 times, (i.e., see to it that the stocks in each of the ten subportfolio have zero correlation with the stocks in the other portfolios). Here, each of the 20 stocks has an initial investment of $500. In Figure 6.6, we show the profile of 500 simulations of such a portfolio. The standard deviation of the pooled fund is only 1487.

How realistic is it to find uncorrelated subportfolios with stocks in each portfolio negatively correlated? Not very. We note that we can increase the sizes of the subportfolios if we wish, only remembering that we cannot pick an arbitrary correlation matrix—it must be positive definite. If we have a subportfolio of k stocks, then if the stocks are all to be equally negatively correlated, the maximum absolute value of the correlation is given by $1/(k-1)$.

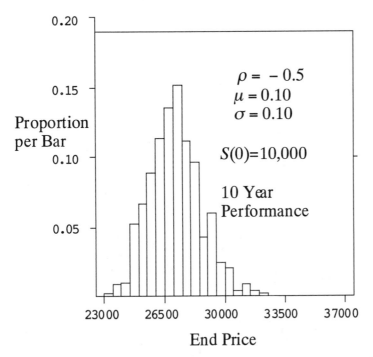

$$\rho = -0.5$$
$$\mu = 0.10$$
$$\sigma = 0.10$$

$$S(0) = 10,000$$

10 Year
Performance

Figure 6.6. Histogram of 10 Independent Two-Stock Portfolios with $\rho = -.5$.

Let us consider another type of randomness in the stock market. Super-imposed over Brownian geometric motion of stocks there are periodic bear market downturns in the market overall. It is unusual for bull markets to exhibit sharp sudden rises. But 10% corrections (i.e., rapid declines) are quite common, historically averaging a monthly probability of as much as .08. Really major downturns, say 20%, happen rather less frequently, say with a monthly probability of .015.

6.6 Bear Jumps

In Figure 6.7 we see the histogram of 500 simulations with the jumps modeled as above, $\sigma = .10$, $\rho = 0$, and $\mu = .235$.

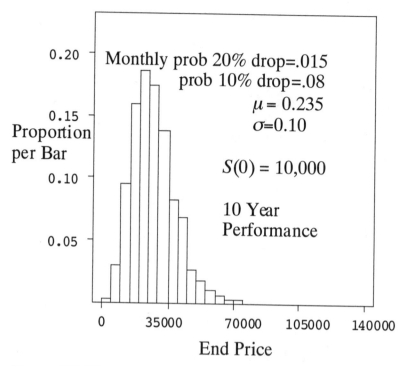

Figure 6.7. Histogram of 20 Independent Stocks with Bear Jumps.

The mean here is 27,080, very similar to that of the situation with independent stocks, with $\mu=.10$ and $\sigma=.1$. However, we note that the standard deviation is a hefty 11,269. We note that these general (across the market) downward jumps take away some of the motivation for finding stocks which have local negative correlation in their movements. (For example, had our portfolio had a .8 correlation between the stochastic increments, the standard deviation would only have increased from 11,269 to 13,522.) Now, we have arrived at a situation where nearly 25% of the time, our portfolio performs worse than a riskless security invested at a 6% return. If we increase the volatility σ to .5, then nearly 40% of the simulated portfolios do worse that the fixed 6% security.

Let us return to looking at the situation where a $100 million university endowment, consisting of 20 stocks ($5 million invested in each stock) with stochastically independent geometric Brownian steps, with $\mu=.235$, $\sigma=.1$ and with monthly probabilies .08 of a 10% drop in all the stocks and a .015 probability of a 20% drop in all the stocks. We shall "spend" the portfolio at the rate of 5% per year. Let us see in Figure 6.8 what the situation might be after 10 years. With probability .08, after 10 years, the endowment will have shrunk to less than $50 million. With probability .22, it will have shrunk to less than $75 million. With probability .34, it will have shrunk to less than the original $100 million. Given that universities tend to spend up to their current cash flow, it is easy to see how some of those which tried such a strategy in the 1960s went through very hard times in the 1970s and 1980s.

It is very likely the case that broad sector downward jumps ought to be included as part of a realistic model of the stock market. The geometric Brownian part may well account for the bull part of the market. But downward movements may very well require a jump component. By simply noting variations on geometric Brownian drift in stock prices, analysts may be missing an essential part of the reality of the stock market, namely large broad market declines which occur very suddenly.

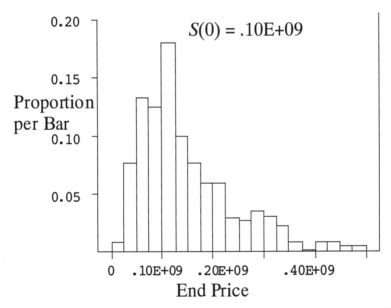

Figure 6.8. Histogram of $100 Million Endowment with 5% Payout.

6.7   Options

In their provocative book *Financial Calculus*, Baxter and Rennie [1] state that for the purpose of pricing derivatives, "the strong law (of large numbers) is completely useless." They show how serious they are about this view when they consider a wager which is based on the progress of a stock. At present, the stock is worth $1. In the next tick of time, it will either move to $2.00 or to $0.50. A wager is offered which will pay $1.00 if the stock goes up and $0.00 if the stock goes down. The authors form a portfolio consisting of 2/3 of a unit of stock and a borrowing of 1/3 of a $1.00

riskless bond. The cost of this portfolio is $0.33 at time zero. After the tick, it will either be worth $2/3 \times \$2.00 - 1/3 \times \$1.00 = \$1.00$ or $2/3 \times \$ 0.50 - 1/3 \times \$1.00 = \$0.00$. From this, they infer that "the portfolio's initial value of $0.33 is also the bet's initial value." In eliminating probabilities from the picture, Baxter and Rennie would pay $0.33 to buy the bet whether the probability of the stock going up were .9999 or .0001. They refer to this example as "the whole story in one step." Of course, the story is patently absurd. Unfortunately, the unmodified version of the Black–Scholes formula does have elements of the Baxter–Rennie one-step. We shall show how sensible modification can eliminate the absurdity.

At any given time, an investor who believes that a stock is going to make serious upward progress over the next six months may purchase an option to buy that stock at any value above its current market listing six months in the future. Such an option is called a *European option*. The price at which he may buy the stock is called the *exercise price* or *strike price* and the time at which the option can be exercised is called the *expiration date*. The option need not be exercised and will not be unless the value of the stock at the expiration date is at least as great as the *exercise price*. *American options* allow the buyer to acquire the stock at any time before the expiration date. Let us suppose that the stock price today is $100. We purchase an option to buy in six months for the strike price of $110. We shall assume, as we stated at the beginning of the chapter, that the stock does not pay dividends. Suppose we have an American option with an expiration date of six months from today, and suppose that, if we acquire the stock at any time during the option period, we intend to keep it at least until the balance of the six month period is over. Suppose the stock goes to $115 three months from today. Should we exercise the option or wait until the full expiration date of six months? If we buy the stock for $110, we will have made a profit of $5 less the price we paid for the option. But if the stock then goes down to $105 by the end of the six months, we will have, at that time incurred a loss of $10 plus the cost of the option. We would have been better off not to have exercised the option at the three month time.

If the stock, on the other hand increases to $125 by the end of the six months, we would have been better off again not to have purchased at three months, for then we would have lost interest on our money for three months. Clearly, then, without dividends, American and European options for the longer term investor have essentially the same values.

What should be the fair value for an option to purchase a stock at an exercise price X starting with today's stock price $S(0)$ and an expiration time of T? If the rate of growth of the stock is μ, and the volatility is σ, then, assuming the exponential Brownian model, the stock value at the time T should be

$$S(0) \exp\left(\mathcal{N}\left((\mu - \frac{\sigma^2}{2})t, t\sigma^2\right)\right),$$

where $\mathcal{N}(a, b)$ is a Gaussian random variable with expectation a and variance b. If we borrow money at a fixed riskless rate r to purchase the option, then the value of the option could be argued to be equal to the

Method A $C_A = \exp(-rT)E[\text{Max}(0, S(T) - X)],$

where E denotes expectation.

On the other hand, it could also be argued that the person buying the option out of his assets is incurring an opportunity cost by using money to buy the option which might as well have been used for purchasing the stock so that the value of the option should be given by

Method B $C_B = \exp(-\mu T) \, E[\text{Max}[0, S(T) - X]].$

Interestingly, in 1971, Black and Scholes [2] came up with a dramatically different strategy. In (6.6), $a = \mu S$ and $b = \sigma S$. Let f be a derivative security (i.e., one which is contingent on S). Then, from Ito's lemma, we have:

$$df = \left(\frac{\partial f}{\partial S}\mu S + \frac{\partial f}{\partial t} + \frac{1}{2}\frac{\partial^2 f}{\partial S^2}(\sigma S)^2 \right) dt + \frac{\partial f}{\partial S}\sigma S dz. \qquad (6.26)$$

Multiplying (6.12) by $\partial f/\partial S$, and isolating $\partial f/\partial S \sigma S dz$ on the left side in both (6.12) and (6.26), we have:

$$\frac{\partial f}{\partial S}\sigma S dz = \frac{\partial f}{\partial S}dS - \frac{\partial f}{\partial S}\mu S dt$$

$$\frac{\partial f}{\partial S}\sigma S dz = df - \left(\frac{\partial f}{\partial S}\mu S + \frac{\partial f}{\partial t} + \frac{1}{2}\frac{\partial^2 f}{\partial S^2}(\sigma S)^2 \right) dt.$$

Setting the two right-hand sides equal (!), we have:

$$df - \frac{\partial f}{\partial S}dS = \left(\frac{\partial f}{\partial t} + \frac{1}{2}\frac{\partial^2 f}{\partial S^2}(\sigma S)^2 \right) dt. \qquad (6.27)$$

Let us consider a portfolio which consists of one unit of the derivative security and $-\partial f/\partial S$ units of the stock. The instantaneous value of the portfolio is then

$$\mathcal{P} = f - \frac{\partial f}{\partial S}S. \qquad (6.28)$$

Over a short interval of time, the change in the value of the portfolio is given by

$$d\mathcal{P} = df - \frac{\partial f}{\partial S}dS = \left(\frac{\partial f}{\partial t} + \frac{1}{2}\frac{\partial^2 f}{\partial S^2}(\sigma S)^2 \right) dt. \qquad (6.29)$$

Now since (6.28) has no dz term, the portfolio is riskless during the time interval dt. We note that the portfolio consists in buying both an option and selling the stock. Since, over an infinitesimal time interval, the Black–Scholes portfolio is a riskless hedge, it could be argued that the portfolio should pay at the rate r of a risk free security, such as a Treasury short term bill. That means that

$$dP = rPdt = r\left(f - \frac{\partial f}{\partial S}S\right)dt = \left(\frac{\partial f}{\partial t} + \frac{1}{2}\frac{\partial^2 f}{\partial S^2}(\sigma S)^2\right)dt. \qquad (6.30)$$

Finally, that gives us the Black–Scholes differential equation

$$rf = \left(rS\frac{\partial f}{\partial S} + \frac{\partial f}{\partial t} + \frac{1}{2}\frac{\partial^2 f}{\partial S^2}(\sigma S)^2\right). \qquad (6.31)$$

It is rather amazing that the Black–Scholes formulation has eliminated both the Wiener term and the stock growth factor μ. Interestingly, however, the stock's *volatility* σ remains. Essentially, the Black–Scholes evaluation of a stock is simply driven by its volatility, with high volatility being prized. We note that μ has been replaced by the growth rate r of a riskless security. Over a short period of time, the portfolio will be riskless. (We recall how, in the Black–Scholes solution, we used a hedge where we bought options and sold stock simultaneously.) This risklessness will not be maintained at the level of noninfinitesimal time. However, if one readjusts the portfolio, say, daily, then (making the huge assumption that sudden jumps cannot happen overnight), it could be argued that assuming one knew the current values of r and σ, a profit could be obtained by purchasing options when the market value was below the Black–Scholes valuation and selling them when the market value was above that of the Black–Scholes valuation (assuming no transaction costs). (Such a fact, it could be argued, in which all traders acted on the Black–Scholes valuation, would drive the market. And that, apparently, is largely the case.)

Now, we recall that a *European call option* is an instrument which gives the owner the right to purchase a share of stock at the *exercise price* X, T time units from the date of purchase. Naturally, should the stock actually be priced less than X at time T, the bearer will not exercise the option to buy at rate X. Although we get to exercise the option only at time T, we must pay for it today. Hence, we must discount the value of an option by the factor $\exp(-rt)$. Since we have seen that the Black–Scholes equation involves no noise term, it is tempting to conjecture that the fair evaluation of an option to purchase a share of stock at exercise price X is given by

$$\begin{aligned} C_{BS} &= e^{-rT}E[\text{Max}(S(T) - X, 0)] \qquad &(6.32)\\ &= e^{-rt}\frac{1}{\sqrt{2\pi\sigma^2 t}}\int_{\ln[\frac{X}{S(0)}]}^{\infty}(S(0)e^z - X)\exp\left[-\frac{1}{2\sigma^2 t}(z - (r - \frac{\sigma^2}{2})t)^2\right]dz \end{aligned}$$

where we note that the growth rate is r rather than μ.

The evaluation of each of the option prices considered above is made rather easy if one uses the following lemma ([5], p. 17).

Lemma 6.1. If S is lognormal with growth rate μ and volatility σ and if

$$
\begin{aligned}
Q &= \lambda S - \gamma X \text{ if } S - \psi X \geq 0 \\
&= 0 \qquad \text{if } S - \psi X < 0,
\end{aligned}
$$

then

$$
\begin{aligned}
E(Q) &= \int_{\psi X}^{\infty} (\lambda S - \gamma X) f(S) dS \\
&= e^{\mu T} \lambda S(0) \Phi \left(\frac{\log(S(0)/X) - \log(\psi) + [\mu + (\sigma^2/2)]T}{\sigma\sqrt{T}} \right) \\
&\quad - \gamma X \Phi \left(\frac{\log(S(0)/X) - \log(\psi) + [\mu - (\sigma^2/2)]T}{\sigma\sqrt{T}} \right), \quad (6.33)
\end{aligned}
$$

where λ, γ, and ψ are arbitrary parameters and Φ is the standard Gaussian cumulative distribution function.

Then we have for **Method A**, taking $\psi=1$ and $\lambda = \gamma = e^{-rT}$,

$$
\begin{aligned}
C_A &= e^{-rT} \{ e^{\mu T} S(0) \Phi \left(\frac{\log(S(0)/X) + [\mu + (\sigma^2/2)]T}{\sigma\sqrt{T}} \right) \\
&\quad - X \Phi \left(\frac{\log(S(0)/X) + [\mu - (\sigma^2/2)]T}{\sigma\sqrt{T}} \right) \}. \qquad (6.34)
\end{aligned}
$$

For **Method B**, taking $\psi = 1$ and $\lambda = \gamma = e^{-\mu T}$,

$$
\begin{aligned}
C_B &= e^{-\mu T} \{ e^{\mu T} S(0) \Phi \left(\frac{\log(S(0)/X) + [\mu + (\sigma^2/2)]T}{\sigma\sqrt{T}} \right) \\
&\quad - X \Phi \left(\frac{\log(S(0)/X) + [\mu - (\sigma^2/2)]T}{\sigma\sqrt{T}} \right) \}. \qquad (6.35)
\end{aligned}
$$

For **Black–Scholes**, taking $\psi = 1$, $\lambda = \gamma = e^{-rT}$, and setting $\mu = r$,

$$
\begin{aligned}
C_{BS} &= e^{-rT} \{ e^{rT} S(0) \Phi \left(\frac{\log(S(0)/X) + [r + (\sigma^2/2)]T}{\sigma\sqrt{T}} \right) \quad (6.36) \\
&\quad - X \Phi \left(\frac{\log(S(0)/X) + [r - (\sigma^2/2)]T}{\sigma\sqrt{T}} \right) \}.
\end{aligned}
$$

Smith [5] gives numerous studies which indicate that a strategy of buying options selling for less than the Black–Scholes value and selling those which exceed the Black–Scholes value produces, on average, a profit. Black and Scholes found that the transaction costs to a private investor outweighed the average profit and therefore inferred that the market as it performed prior to the use of the Black–Scholes model was efficient. The Black–Scholes model has had enormous impact on the trading of options. Consequently, it has itself changed the mechanism of the market.

Consider, in Tables 6.2 and 6.3, the performance of the Black–Scholes model compared to Model A and Model B in the case where a stock has growth rate $\mu = .15$ with a fixed riskless interest rate of 5% and a variety of volatilities. We shall assume the option is for an exercise time of six months in the future, and that the price of the stock at the present time is $100. A variety of exercise prices is considered; 20,000 simulations are used.

Table 6.2. Six-Month Options: $\sigma = .20$, $\mu = .15$									
Ex Price	102	104	106	108	110	112	114	116	118	120
C_{BS}	5.89	4.99	4.20	3.51	2.91	2.40	1.98	1.63	1.36	1.16
C_B	8.58	7.47	6.46	5.55	4.73	4.00	3.37	2.82	2.35	1.95
C_A	9.02	7.85	6.79	5.83	4.97	4.21	3.54	2.96	2.47	2.05
Sim Av	8.97	7.79	6.72	5.76	4.89	4.14	3.47	2.89	2.40	1.97

Table 6.3. Six-Month Options: $\sigma = .40$, $\mu = .15$									
EP	102	104	106	108	110	112	114	116	118	120
C_{BS}	11.48	10.63	9.82	9.07	8.37	7.72	7.11	6.54	6.01	5.53
C_B	13.84	12.89	12.00	11.16	10.37	9.62	8.92	8.26	7.64	7.06
C_A	14.55	13.55	12.62	11.73	10.90	10.11	9.37	8.68	8.03	7.43
SA	14.39	13.39	12.45	11.56	10.73	9.94	9.21	8.52	7.87	7.27

An increase in volatility (σ) typically increases the value of the option. We also note that the simulation average (time discounted using an interest rate of .05) is most closely approximated by C_A and not very well at all approximated by the Black–Scholes price. This should be hardly surprising, since the simulation average is essentially a Monte Carlo approximation to C_A. C_{BS} appears to undervalue an option. But the fact remains that if our assumptions of geometric Brownian drift, and so on, are correct, then, on the average, C_A gives the expected break-even price to a buyer who pays for his options with money borrowed at 5% interest. Clearly, we need to investigate the apparent discrepancies between C_{BS} and, say, C_A, further.

Consider the performance of the Black–Scholes model compared to Model A and Model B when a stock has growth rate $\mu = .15$ with a fixed riskless interest rate of 5% and a variety of volatilities. Assume that the option is for an exercise time of six months in the future, and that the price of the stock at the present time is $100. Several exercise prices are considered in Tables 6.4a–d.

Table 6.4a. Six-Month Options: $\sigma = .001$ $\mu = .15$									
Ex Price	102	104	106	108	110	112	114	116	118	120
C_{BS}	0.52	0.00	0.00	0.00	0.00	0.00	0.00	0.00	0.00	0.00
C_B	5.37	3.51	1.66	0.00	0.00	0.00	0.00	0.00	0.00	0.00
C_A	5.65	3.69	1.74	0.00	0.00	0.00	0.00	0.00	0.00	0.00

Table 6.4b. Six-Month Options: $\sigma = .20$ $\mu = .15$									
Ex Price	102	104	106	108	110	112	114	116	118	120
C_{BS}	5.89	4.99	4.20	3.51	2.91	2.40	1.98	1.63	1.36	1.16
C_B	8.58	7.47	6.46	5.55	4.73	4.00	3.37	2.82	2.35	1.95
C_A	9.02	7.85	6.79	5.83	4.97	4.21	3.54	2.96	2.47	2.05

			Table 6.4c. Six-Month Options: $\sigma = .40\ \mu = .15$							
EP	102	104	106	108	110	112	114	116	118	120
C_{BS}	11.48	10.63	9.82	9.07	8.37	7.72	7.11	6.54	6.01	5.53
C_B	13.84	12.89	12.00	11.16	10.37	9.62	8.92	8.26	7.64	7.06
C_A	14.55	13.55	12.62	11.73	10.90	10.11	9.37	8.68	8.03	7.43

			Table 6.4d. Six-Month Options: $\sigma = 2.00\ \mu = .15.$							
EP	102	104	106	108	110	112	114	116	118	120
C_{BS}	52.17	51.71	51.25	50.80	50.36	49.93	49.50	49.08	48.67	48.26
C_B	53.37	52.91	52.45	52.00	51.56	51.13	50.70	50.28	49.87	49.47
C_A	56.11	55.62	55.14	54.67	54.21	53.75	53.30	52.86	52.43	52.00

Again, we note that an increase in volatility (σ) typically increases the value of the option. Next, let suppose that the stock is actually declining in value. What happens when, say, $\mu = -.15$ is indicated in Tables 6.5a–d.

			Table 6.5a. Six-Month Options: $\sigma = .001\ \mu = -.15$							
Ex. Price	102	104	106	108	110	112	114	116	118	120
C_{BS}	0.52	0.00	0.00	0.00	0.00	0.00	0.00	0.00	0.00	0.00
C_B	0.00	0.00	0.00	0.00	0.00	0.00	0.00	0.00	0.00	0.00
C_A	0.00	0.00	0.00	0.00	0.00	0.00	0.00	0.00	0.00	0.00

			Table 6.5b. Six-Month Options: $\sigma = .20\ \mu = -.15$							
Ex. Price	102	104	106	108	110	112	114	116	118	120
C_{BS}	5.89	4.99	4.20	3.51	2.91	2.39	1.95	1.59	1.28	1.02
C_B	2.22	1.77	1.40	1.10	0.85	0.66	0.51	0.38	0.29	0.22
C_A	2.01	1.60	1.27	0.99	0.77	0.60	0.46	0.35	0.26	0.20

			Table 6.5c. Six-Month Options: $\sigma = .40\ \mu = -.15$							
Ex. Pr	102	104	106	108	110	112	114	116	118	120
C_{BS}	11.48	10.63	9.82	9.07	8.37	7.72	7.11	6.54	6.01	5.53
C_B	7.49	6.83	6.22	5.66	5.15	4.68	4.24	3.85	3.48	3.15
C_A	6.77	6.18	5.63	5.12	4.66	4.23	3.84	3.48	3.15	2.85

			Table 6.5d. Six-Month Options: $\sigma = 2.00\ \mu = -.15$							
EP	102	104	106	108	110	112	114	116	118	120
C_{BS}	52.17	51.71	51.25	50.80	50.36	49.93	49.50	49.08	48.67	48.26
C_B	49.77	49.30	48.84	48.39	47.95	47.51	47.08	46.66	46.25	45.84
C_A	45.03	44.61	44.19	43.78	43.38	42.99	42.60	42.22	41.85	41.48

We note that Black–Scholes values a call option at, say $110, equally whether the growth rate of the stock is +.15 or -.15. This appears a bit bizarre.

Let us consider limiting behavior as the volatility goes first to infinity and then to zero. Suppose that a stock is currently selling for $S(0)$. We wish to buy an option T time units in the future with strike price X. As the volatility of the stock goes to infinity, then we note that both Black–Scholes and Method B tell us that the option is so valuable that its fair price is simply the current value of the stock, namely $S(0)$, irrespective of the value of μ.

On the other hand, let us suppose that the value of the volatility is zero. Then the Black–Scholes price is

$$
\begin{aligned}
C_{BS} &= S(0) - e^{-rT}X \text{ if } S(0)e^{rT} \geq X \\
&= 0 \text{ otherwise.}
\end{aligned}
\tag{6.37}
$$

Consider the following intuitive argument supportive of the Black–Scholes price. Typically, the stock will grow at a rate μ greater than the riskless rate r. Since the process has no volatility, there is no uncertainty about it. Either the stock grows above the strike price or it does not, and we know the result *a priori* . If it does not, then the option is worthless. If the stock grows above the strike price, then the value will be worth the difference of the stock price and the strike price. So, the company that sells the option contract can buy the stock now at price $S(0)$. To buy the stock, the vendor borrows $S(0)$ dollars at the riskless interest rate r. By the time that the option is to be exercised, the cost to the vendor is $S(0)e^{rT}$. The buyer, at time zero, puts up the price C_{BS}. So then, by time T, the vendor has a stock worth $S(T)$ and the "bond," which will have grown in value to $C_{BS}e^{rT}$. If the value at time T is greater than X, then the buyer will exercise his or her option, and pay the vendor X dollars for the bond. In that case, in terms of dollar values at time T, the vendor will sell the bond for X dollars. So, at time T he will have assets equal to $-S(0)e^{rt} + X + C_{BS}e^{rT}$. Taking this back to time zero, the value of assets to the vendor will be $-S(0) + Xe^{-rT} + C_{BS}$. This can be made equal to zero only if $C_{BS} = S(0) - Xe^{-rT}$. (As we have stated, μ is typically greater than the riskless rate r. So, if $S(0)e^{rT} \geq X$, then $S(0)e^{\mu T} > X$.) The classical Black–Scholes dogma would tell us that the vendor should sell the option for $C_{BS} = S(0) - Xe^{-rT}$ plus a commission.

There is a practical problem with this argument. If the volatility is zero and the growth rate is $\mu > r$, the stock is behaving like a riskless security paying at a rate greater than the standard r. Why would anyone lend us money at rate r under such circumstances? Clearly, they would prefer to buy the stock themselves.

Of course, the standard neoclassical argument to this point is that if the stock has no volatility, its price would not be in equilibrium at the point of the rate of the option. Hence, the price would have to rise to cause $\mu = r$, and the stock would be no different from a Treasury bill. This would, in fact, happen in the neoclassical world of perfectly efficient markets where prices are always in equilibrium. However, in the real world, where market imperfections make it possible for a company to earn economic rent, it probably does not make sense for the financial markets to clear at competitive prices while all other markets (factor inputs, goods, etc.) do not. This point was shown a number of years ago by Williams and Findlay [7]. Nevertheless, neoclassical models regularly appear suggesting that companies <u>can</u> have growth rates where $\mu > r$, even though under perfect market conditions, all future rents should have been discounted completely!

More generally, in the case where there is a positive volatility in the stock, the lender would still face the fact that if one could find a large enough portfolio of not perfectly correlated securities growing at an average rate of $\eta > r$, then it would be foolish to lend out money at rate $r < \mu$ when it could rather be invested in the portfolio. The vendor of the option typically has no strong views about a particular stock. He or she is selling options in many stocks and is only interested that he or she retrieves his or her supposed opportunity cost rate η plus a commission. Accordingly, the vendor of the option will expect to use Black–Scholes with r replaced by η (plus, of course, a commission).

$$
C_{\text{vendor}} = e^{-\eta T}\{e^{\eta T}S(0)\Phi\left(\frac{\log(S(0)/X) + [\eta + (\sigma^2/2)]T}{\sigma\sqrt{T}}\right)
$$
$$
-X\Phi\left(\frac{\log(S(0)/X) + [\eta - (\sigma^2/2)]T}{\sigma\sqrt{T}}\right)\}. \tag{6.38}
$$

On the other hand, the buyer of the option will have fairly strong views about the stock and its upside potential. The buyer should use Black–Scholes replacing r by μ, where, typically, $\mu > \eta > r$. Thus,

$$
C_{\text{buyer}} = e^{-\mu T}\{e^{\mu T}S(0)\Phi\left(\frac{\log(S(0)/X) + [\mu + (\sigma^2/2)]T}{\sigma\sqrt{T}}\right)
$$
$$
-X\Phi\left(\frac{\log(S(0)/X) + [\mu - (\sigma^2/2)]T}{\sigma\sqrt{T}}\right)\}. \tag{6.39}
$$

We note that typically, in the mind of the call option buyer, μ is rather large. Perhaps some investors will bet solely on the basis of a large stock volatility, but this is unusual. Option buying is frequently a leveraging device whereby an investor can realize a very large gain by buying call options rather than stocks. The seller of the option is probably expecting an $\eta < \mu$ value as the reasonable rate of return on its investments overall. It is observed [6] that the arithmetic mean annual return on U.S. common stocks from 1926 on is over 10%. Let us suppose we are dealing with an initial stock price of $100 and that the vendor uses $\eta = .10$ and the buyer believes $\mu = .15$. In Tables 6.6a and b, we show the values of $C_{BS(\text{vendor})}$ and $C_{BS(\text{buyer})}$, respectively. This may appear confusing, for we have arrived at a Black–Scholes price for the vendor and one for the buyer, and they are generally not the same. *Pareto optimality* is the situation where all parties are better off by undertaking a transaction. Clearly, at least from their respective viewpoints, we do have Pareto optimality (assuming that the commission is not so high as to swamp the anticipated profit to the buyer). It is true that from the standpoint of the seller of an option, break-even may be good enough, for there is profit to be made from the commission. But break-even is not sufficient for the buyer. He or she must expect that there is a profit to be made from the transaction.

Table 6.6a. Six-Month Options. $\sigma = .20$										
Ex Price	102	104	106	108	110	112	114	116	118	120
C_{vendor}	7.17	6.16	5.26	4.45	3.74	3.13	2.59	2.13	1.75	1.42
C_{buyer}	8.58	7.47	6.46	5.55	4.73	4.00	3.37	2.81	2.33	1.92

Table 6.6b. Six-Month Options. $\sigma = .40$										
Ex Pr	102	104	106	108	110	112	114	116	118	120
C_{ven}	12.63	11.73	10.88	10.08	9.34	8.63	7.98	7.36	6.79	6.25
C_{buy}	13.84	12.89	12.00	11.16	10.36	9.62	8.91	8.25	7.64	7.06

Let us address the question of how options are actually bought and sold. A large pension fund may decide to generate income by selling options in a stock. The fund will, typically, demand a return on investment which is comparable to that of the portfolio it manages. Other vending institutions will generally take the same view. It will have effect on the market. The naive Black–Scholes formula, using the riskless rate r, say, of Treasury bills, may be useful as a floor for the option price, but some grand average of portfolio returns, say η, should be investigated as well. On the other hand, the options will be sold to individuals who expect a high rate of return, say $\mu > \eta > r$, and will be willing to pay a Black–Scholes rate with the seller's η which will be greater than r but less than μ.

6.8   Getting Real: Simulation Analysis of Option Buying

Rather than defaulting to formulas, the wise investor probably would like to decide whether or not the "asking price" for an option is attractive based on his or her informed opinion as to what will happen to the underlying stock. The seller of the option does not have the same position as that of the investor. The "asking price" will use Black–Scholes based on some broad rate of return, say η (probably not that of a Treasury bill, probably something rather higher, plus a commission). It may well reflect the volatility of a particular stock, but is unlikely to reflect the buyer's optimism in terms of the future growth of the stock.

Once we leave the realm of standard Brownian models, we generally do not have ready formulas available to answer questions as to the wisdom of buying an option or not. Simulation gives us a way to cope whenever we have a reasonable notion, even a nonstandard notion, as to what is going on. For example, suppose that we wished to superimpose on the geometric Brownian walk a bear jump which discounts the value of the stock by 10% on any given month with probability .08 and the value of the stock by 20%

with probability .015 on any given month. This would represent roughly an anticipated bear downturn of 14%.

Table 6.7a. Six-Month Options (with Bear Jumps). $\sigma = .20$, $\mu = .15$										
Ex. Pr.	102	104	106	108	110	112	114	116	118	120
C_{BS}	5.89	4.99	4.20	3.51	2.91	2.40	1.98	1.63	1.36	1.16
C_B	8.58	7.47	6.46	5.55	4.73	4.00	3.37	2.82	2.35	1.95
C_A	9.02	7.85	6.79	5.83	4.97	4.21	3.54	2.96	2.47	2.05
Sim.	5.97	5.13	4.38	3.72	3.13	2.62	2.18	1.80	1.48	1.21

Table 6.7b. Six-Month Options (with Bear Jumps). $\sigma = .40$, $\mu = .15$										
EP	102	104	106	108	110	112	114	116	118	120
C_{BS}	11.48	10.63	9.82	9.07	8.37	7.72	7.11	6.54	6.01	5.53
C_B	13.84	12.89	12.00	11.16	10.37	9.62	8.92	8.26	7.64	7.06
C_A	14.55	13.55	12.62	11.73	10.90	10.11	9.37	8.68	8.03	7.43
Sim.	10.89	10.08	9.33	8.62	7.96	7.34	6.76	6.23	5.73	5.28

Now, based on Tables 6.7a and b, the pricing of the Black–Scholes model appears inspired. Of course, we have simply added on the kind of unexpected downward turn which is not accounted for by the geometric Brownian walk unmodified. On the other hand, our imposition of bear jumps has depressed the expected growth rate of the stock to essentially 1%, and most of the value of the option is due to volatility.

Suppose that an investor believes the rate of growth of a stock is .15 overall, bear jumps included. Then, if we are to include the bear jumps, we need to increase the value of the Brownian growth to $.15 + .14 = .29$. So, let us now compute the simulated buyer's price, with discount to present value rate being $\mu = .15$. We also compute the vendor's price using the Black–Scholes formula with riskless rate $\eta = .10$ (we will assume that the vendor will use the nominal volatility values of .20 and .40, as shown in Tables 6.8a and b).

Table 6.8a. Six-Month Options (with Bear Jumps): $\sigma = .20$, $\mu = .15$										
Ex Pr	102	104	106	108	110	112	114	116	118	120
C_{BS}	5.89	4.99	4.20	3.51	2.91	2.40	1.98	1.63	1.36	1.16
C_{vendor}	7.17	6.16	5.26	4.45	3.74	3.13	2.59	2.13	1.75	1.42
C_{buyer}	10.53	9.34	8.24	7.23	6.30	5.46	4.71	4.04	3.44	2.91

Table 6.8b. Six-Month Options (with Bear Jumps): $\sigma = .40$, $\mu = .15$										
EP	102	104	106	108	110	112	114	116	118	120
C_{BS}	11.48	10.63	9.82	9.07	8.37	7.72	7.11	6.54	6.01	5.53
C_{ven}	12.63	11.73	10.88	10.08	9.34	8.63	7.98	7.36	6.79	6.25
C_{buy}	15.56	14.54	13.58	12.67	11.80	10.98	10.21	9.49	8.81	8.17

It is unlikely that a buyer will be able to acquire options at the orthodox Black–Scholes rate (i.e., the one using Treasury bill interest rates of .05). However, even if the seller is using Black–Scholes with the more realistic .10, it would appear that the buyer values the option at a value higher than the vendor rate in each of the cases shown above. She might, accordingly, be tempted to purchase, say, an option to buy at a strike price of $108.

Simulation allows her to do more. She can obtain a profile of the anticipated values of such an option. In Figure 6.9, we note, for $\sigma = .2$, that the value of the option will be zero around 55% of the time. The expected value to the buyer will be $7.23, but if she wishes to purchase the option, she should realize that around 55% of the time she will have lost her purchase price of the option. There are many other things the prospective buyer might choose to try before making the decision as to whether or not the option should be bought. Most of these are rather easy to achieve with simulation.

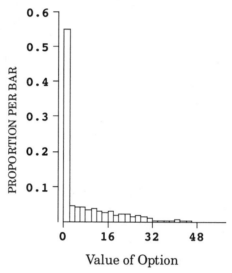

Value of Option

Figure 6.9. Histogram of Option (Present) Values.

6.9   Conclusions

Persons who indulge in market purchases based on fine notions of a Brownian regularity and rapidly attained equilibria unmoderated by the modifications which are almost always necessary if we are to look at real stocks from the varying standpoints of real people and real institutions, should not be surprised when they lose their (or more commonly other people's) shirts. A recent example [4] is the debacle of the the Long-Term Capital Management Limited Partnership (LTCM, LP). The fund was conceived and managed by John Meriwether, a former Salomon Brothers vice chairman with advice from Myron Scholes and Robert Merton, the 1997 winners of the Nobel Prize in Economics for developing the Black–Scholes options pricing model with the late Fischer Black. Because the enormous losses of the LTCM might well have caused a chain reaction leading to a market crash, Federal Reserve Chairman Allan Greenspan, in an almost unprecedented move, put together a $3.5 billion bailout. The portfolio, at the time of the bailout, was priced at two cents on the dollar.

One of Benjamin Franklin's maxims is "Experience keeps a hard school, and a fool will learn by no other." In the two decades between the publication of the Black–Scholes equation and the formation of the LTCM, it might have been better if investigators had stressed the model by massive simulations, using both historical data and "what if?" scenarios. The cost would have been perhaps a few million dollars for salaries and computer time. Now, by actual utilization of real money in a real market, we have learned that the assumptions of the Black–Scholes options pricing model are flawed. But at enormous cost.

The whole area of modeling markets is an exciting one. Anyone investing or anyone advising an investor may make ruinous mistakes if they do not ask questions such as "what happens if ...". Such questions, typically, are not easy to answer if we demand closed-form solutions. Simulation enables us to ask the questions we need to ask rather than restricting ourselves to looking at possibly irrelevant questions for which we have formulas at hand.

Problems

6.1. Let X be a random variable with the property that $Y = \ln X$ has a normal distribution with mean μ and variance σ^2. Prove that:
(a) The mean of X is $\exp(\mu + \sigma^2/2)$.
(b) The variance of X is $[\exp(\sigma^2) - 1] \exp(2\mu + \sigma^2)$.

6.2. Consider a portfolio of n stocks, each with starting value $S(0)$ and the same μ and σ. Suppose for each stock

$$\frac{\Delta S_i}{S_i} = \mu \Delta t + \sigma \epsilon_i \sqrt{\Delta t}$$

$$\epsilon_i = c(a\eta_0 + (1-a)\eta_i)$$

$$a = \frac{\rho - \sqrt{\rho(1-\rho)}}{2\rho - 1}$$

and

$$c^2 = \frac{1}{a^2 + (1-a)^2}.$$

Show that the variance of the portfolio's value at time t is given by

$$\text{Var} = n^2 S(0)^2 e^{(2\mu t)}[\exp(\rho \sigma^2 t) - 1] + n S(0)^2 e^{(2\mu t)}[\exp(\sigma^2 t) - \exp(\rho \sigma^2 t)].$$

6.3. Let us consider two investors.

(a) Mr. A is considering an "index fund." This will be essentially a mutual fund of stocks randomly selected from a broad category. For example, we could form an index fund of computer-related stocks. Or, we could form an index fund of stocks selected from the American Stock Exchange. The stocks might be purchased in arbitrary amounts or weighted according to the total market value of each particular firm in relation to that of the total value of all stocks on the exchange. A person who chooses to buy stocks in this way is, in effect, assuming that all other purchasers of stocks are behaving rationally (i.e., they are willing to pay more for promising stocks than for stocks seemingly on the downward track). Accordingly, the index fund manager should be perfectly happy to buy in random fashion; he needs no market analysis: the rational behavior of the other players in the market implicitly guides him. He does not worry about any idea of playing a zero-sum game. The technological improvements, present in whatever group that is being indexed, tend to move the value of the index higher.

(b) Ms. B thinks A's strategy is rather dim. She picks the same category as A—say, computer-related stocks. Then she looks at the fundamentals of her choices and invests her capital in a portfolio of stocks where the growth rates and other fundamentals appear to be most favorable.

Which strategy do you favor? Give your reasons.

6.4. Write a simulation program for a single stock with specified growth and volatility. Note that for many traverses of a loop, older languages such as Fortran and C tend to perform much more quickly than more modern user-friendly programs.

6.5. Write a simulation program for a portfolio of k stocks, correlated with each other with correlation equal to ρ but with arbitrary growth rates and volatilities for each stock.

6.6. Take the program in Problem 6.5 and add on the possibility of bear jumps across the market.

6.7. Verify Lemma 6.1 (6.33).

6.8. Consider the Black–Scholes differential equation (6.31) with the boundary condition that

$$\begin{aligned} f(S,T) &= S(T) - X \text{ if } S(T) - X > 0 \\ &= 0, \text{ otherwise.} \end{aligned}$$

Prove that

$$\begin{aligned} f(S,t) &= S\Phi\left(\frac{\log(S/X) + [r + (\sigma^2/2)](T-t)}{\sigma\sqrt{T-t}}\right) \\ &\quad - Xe^{[-r(t-t)]}\Phi\left(\frac{\log(S/X) + [r - (\sigma^2/2)](T-t)}{\sigma\sqrt{T-t}}\right). \end{aligned}$$

6.9. Stocks are tied to real companies, and these companies can get better or worse. Let us consider a mutual fund which is solely managed by a computer. At any given time, there will be 20 stocks in the fund. Let us suppose that these stocks are selected randomly from stocks which, over the last six months, have exhibited a growth rate μ of .25 and a volatility of .15. We will assume, as in the past, that there is an across-the-board bear jump mechanism whereby a sudden drop of 10% happens, on the average, of once a year and a sudden drop of 20% happens on the average once every five years. Let us suppose that there is a mechanism which happens, on the average once every three years, whereby the growth rate of a stock can drop to a level which is uniformly distributed between -.05 and .2 with a corresponding increase in the volatility to a level uniformly distributed between .2 and .4. Since the computer does not read newspapers, it needs to have an automatic mechanism for removing, at the earliest possible time, from the portfolio stocks whose fundamentals have failed in this fashion. When such a stock is removed, it will be replaced by another with growth rate .25 and volatility .15.

(a) Assuming no transaction costs, devise a good automatic scheme for the computer to follow. Show simulation studies to substantiate the wonderfulness of your paradigm.

(b) Assuming transaction costs equal to 1.5% of the value of the stock traded on both the sell and the buy, devise a good strategy for the computer to follow and show its effectiveness by a simulation study.

6.10. A group of investors is considering the possibility of creating a European option-based mutual fund. As a first step in a feasibility study, they decide to compare investments of $1 million in a portfolio of 20 stocks as opposed to a portfolio of 20 options. They want to obtain histograms of investment results after one year. Let us suppose that all stocks in the portfolios are bought at a cost of $100 per share. Let us assume the usual model of stock growth,

$$S(t) = S(0)\exp(\mu t + \sigma\sqrt{t}\epsilon),$$

where ϵ is normally distributed with mean 0 and variance 1. Let us take two values of μ, namely, .10 and .15. Also, let us consider two values of σ, namely, .15 and .30. Consider several strike prices for the options: for example, the expected value of the stock at the end of one year, and various multiples thereof. Assume that the options are completely fungible. Thus, at the end of the year, if a stock is $10 over the strike price, the option purchased for the Black–Scholes price is worth $10 (i.e., one does not have to save capital to buy the stock; one can sell the option). For the "riskless" interest rate, use two values: .06 and .08. Clearly, then, we are considering a leveraged portfolio and seeing its performance in relationship to a traditional one. Carry out the study assuming that there is no correlation between the stocks. Then, see how the situation is changed if one assumes a ρ value in Problem 6.2 of .5.

6.11. Carry out the study in Problem 6.10 with the following modification. Use the μ value of .24 and the two σ values of .15 and .30. Then assume that there is an across-the-board bear jump mechanism whereby a sudden drop of 10% happens, on the average, once a year and a sudden drop of 20% happens on the average once every five years. The overall growth is still roughly .10. Use the Black–Scholes riskless price as before without adding in the effect of the Poisson jumps downward.

References

[1] Baxter, M. and Rennie, A. (1996). *Financial Calculus: An Introduction to Derivative Pricing.* New York: Cambridge University Press.

[2] Black, F. and Scholes, M. (1973). "The pricing of options and corporate liabilities," *J. Political Econ.*, **81**, 637–659.

[3] Hull, J.C., (1993). *Options, Futures, and Other Derivative Securities.* Englewood Cliffs, N.J.: Prentice Hall.

[4] Siconolfi, M., Raghavan, A., and Pacelle, M. (1998). "All bets are off: how the salemanship and brainpower failed at Long-Term Capital," *The Wall Street Journal*, November 16, 1998, A1, A18–A19.

[5] Smith, C.W. (1976). "Option pricing: a review," *J. Financial Econ.*, **3**, 3–51.

[6] Williams, E.E. and Thompson, J.R. (1998). *Entrepreneurship and Productivity.* New York: University Press of America, 87.

[7] Williams, E.E. and Findlay, M.C. (1979). "Capital budgeting, cost of capital, and ex-ante static equilibriium," *J. Business Finance Account.*, Winter, 281–299.

Chapter 7

Simulation Assessment of Multivariate and Robust Procedures in Statistical Process Control

7.1 Introduction

A major use of simulation is the assessment of the performance of estimators and tests in situations where closed form solutions are difficult to obtain. We might come up, based on intuition or some sort of criterion, with an idea for a test of an hypothesis. If we took the time to derive analytical formulas to assess the validity of the procedure, we might very well, at the end of the day, discover, after much labor, that our procedure was not very good. Frequently, we can use simulation as an alternative to the analytical formulas required for the purposes of assessment of performance.

Rather than cover the gamut of situations where such a use of simulation might prove useful, we shall focus on a specific problem area: the performance of some testing and estimation procedures using multivariate data in statistical process control. Generally, the ability to devise simulation-based techniques for performance assessment, developed in one area of statistics, tends to work well more broadly.

In terms of its economic impact, statistical process control is among the most important topics in modern statistics. Although some statisticians (see, e.g., Banks [2]) have considered SPC to be trivial and of scant importance, the market seems to have reacted quite differently. For example, the American Society for Quality is vastly larger in its membership than the

American Statistical Association. It is clear is that SPC is not going away, even should many professional statisticians continue in their disdain for it.

At the end of the World War II, Japan was renowned for shoddy goods produced by automatons living in standards of wretchedness and resignation. W. Edwards Deming began preaching the paradigm of statistical process control (originally advocated by Walter Shewhart) in Japan in the early 1950s. By the mid 1960s, Japan was a serious player in electronics and automobiles. By the 1980s, Japan had taken a dominant position in consumer electronics and, absent tariffs, automobiles. Even in the most sophisticated areas of production, such as computing, the Japanese had achieved a leadership role. The current situation of the Japanese workers is among the best in the world. A miracle, to be sure, and one far beyond that of, say postwar Germany, which was a serious contender in all levels of production before World War II.

There is little doubt that the SPC paradigm facilitated these significant changes in Japanese production. Nevertheless, SPC is based on some very basic notions:

- The key to optimizing the output of a system is the optimization of the system itself.

- Although the problem of modifying the output of a system is frequently one of linear feedback (easy), the problem of optimizing the system itself is one of nonlinear feedback (hard).

- The suboptimalities of a system are frequently caused by a small number of assignable causes. These manifest themselves by intermittent departures of the output from the overall output averages.

- Hence, it is appropriate to dispense with complex methods of system optimization and replace these by human intervention whenever one of these departures is noted.

- Once an assignable cause of suboptimality has been removed, it seldom recurs.

- Thus, we have the indication of an apparently unsophisticated but, in fact, incredibly effective, paradigm of system optimization.

Perhaps there is a valid comparison between Shewhart and Adam Smith, who had perceived the power of the free market. But there appears to be no single implementer of the free market who was as important in validating *The Wealth of Nations* as Deming has been in validating the paradigm of statistical process control. There has never been, in world history, so large scale an experiment to validate a scientific hypothesis as Deming's Japanese validation and extension of the statistical process control paradigm of Deming and Shewhart.

It is not our intention to dwell on the philosophy of SPC. That topic has been extensively dealt with elsewhere (see, e.g., Thompson and Koronacki [9]). We will develop here a modeling framework for SPC and then indicate natural areas for exploration. Both Shewhart and Deming held doctorates in mathematical physics, so it is reasonable to assume that there was some reason they did not resort to exotic mathematical control theory type strategies. In Figure 7.1 we indicate a standard feedback diagram for achieving the desired output of a system.

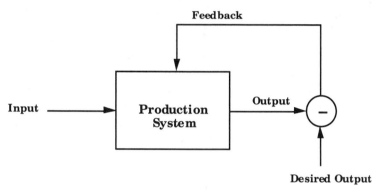

Figure 7.1. Control of Output.

One might pose this as an optimization problem where we desire to minimize, say

$$\int_0^T [\text{output}(t) - \text{target}(t)]^2 \, dt. \tag{7.1}$$

Such a problem is generally linear and tractable.

However, the task of SPC is not to optimize the output directly, but rather, to achieve optimization of the system itself. Generally, such an optimization is an ill-posed problem. At some future time, artificial intelligence and expert systems may bring us to a point where such problems can be handled, to a large extent, automatically. But Shewhart and Deming lacked such software/hardware (as we all still lack it). So they resorted to a piecewise (in time) control strategy based on human intervention (cf. [3] and [5]).

7.2   A Contamination Model for SPC

It is paradoxical that W. Edwards Deming, one of the most important statistical figures of all time, never really published a model of his paradigm of statistical process control (SPC). Deming rightly argued that the key to quality control of an industrial product was to understand the system that produced it. But, in the case of the SPC paradigm itself, he was rather didactic, like the Zen masters of Japan, the country whose economy and standard of living he did so much to improve. Careful analysis of Deming's paradigm led Thompson and Koronacki to their model-based analysis of the SPC system [9].

To understand one of the key aspects of SPC, let us first of all assume that there is a "best of all possible worlds" mechanism at the heart of the process. For example, if we are turning out bolts of 10-cm diameter, we can assume that there will be, in any lot of measurements of diameters, a variable, say X_0, with mean 10 and a variance equal to an acceptably small number. When we actually observe a diameter, however, we may not be seeing only X_0 but a sum of X_0 plus some other variables which are a consequence of flaws in the production process. These are not simply measurement errors but actual parts of the total diameter measurements which depart from the "best of all possible worlds" distribution of diameter as a consequence of imperfections in the production process. One of these imperfections might be excessive lubricant temperature, another bearing vibration, another nonstandard raw materials, and so on. These add-on variables will generally be intermittent in time. This intermittency enables us to find measurements which appear to show "contamination" of the basic production process. We note how different the situation would be without the intermittency, if, say, an output variable were the sum of the "best of all possible worlds" variable X_0 and an "out of control" variable X_1. Then, assuming both variables were Gaussian, the output variable would simply have the distribution $\mathcal{N}(\mu_0 + \mu_1, \sigma_0^2 + \sigma_1^2)$, and the SPC control charts *would not work*. Perhaps the greatest statistical contribution of Shewhart was noting the general presence of intermittent contamination in out-of-control systems.

It is important to remember that the Deming–Shewhart paradigm of SPC is *not* oriented toward the detection of faulty lots. Rather, SPC seeks for atypical lots to be used as an indication of epochs when the system exhibits possibly correctable suboptimalities. If we miss a bad lot or even many bad lots, that is not a serious matter from the standpoint of SPC. If we are dealing with a system that does not produce a sufficiently low proportion of defectives, we should use 100% end inspection, as we note from the following argument:

Let a be the cost of passing a bad item.
Let b be the cost of inspecting an item.
Let x be the proportion of items inspected.
Let y be the proportion of bad items.
Let N be the number of items produced.
Then the cost of inspecting some items and not inspecting others is given by

$$\begin{aligned} C &= bxN + ay(1-x)N \\ &= (b - ay)xN + ayN. \end{aligned}$$

Clearly, then, if $ay > b$, we should inspect all the lots. (If $ay < b$, we should inspect none.)

End product inspection is really not SPC but *quality assurance*. Most companies use some sort of quality assurance. SPC, however, is different from quality assurance. In fact, the experience of Thompson and Koronacki when implementing SPC in a number of factories in Poland was that it was better to leave the quality assurance (a.k.a. quality control) groups in place and simply build up new SPC departments.

Now, for SPC, we simply cannot have the alarms constantly ringing, or we shall be wasting our time with false alarms. Accordingly, in SPC, we should be interested in keeping the probability of a Type I error small. Accordingly, the testing rules in control charting are typically of the type

$$P[\text{declaring "out of control"} \mid \text{in control}] = .002. \qquad (7.2)$$

With such conservatism we may well find an out-of-control situation later rather than sooner. However, we shall tend to avoid situations where the alarms are always ringing, frequently to no good purpose. And by the theory of SPC, suboptimalities, if missed, will occur in the same mode again.

Proceeding with the contamination model, let us assume that the random variables are added. In any lot, indexed by the time t of sampling, we will assume that the measured variable can be written as

$$Y(t) = X_0 + \sum_{i=1}^{k} I_i(t)X_i, \qquad (7.3)$$

where X_i comes from distribution F_i having mean μ_i and variance σ_i^2, $i = 1,2,\ldots, k$. and indicator

$$\begin{aligned} I_i(t) &= 1 \text{ with probability } p_i \\ &= 0 \text{ with probability } 1 - p_i. \end{aligned} \qquad (7.4)$$

When such a model is appropriate, then, with k assignable causes, there may be in any lot, 2^k possible combinations of random variables contributing to Y. Not only do we assume that the observations within a lot are independent and identically distributed, we assume that there is sufficient temporal separation from lot to lot that the parameters driving the Y process are independent from lot to lot. Also, we assume that an indicator variable I_i maintains its value (0 or 1) throughout a lot. Let $\mathcal{I}$ be a collection from $i \in 1, 2, \ldots, k$. Then

$$Y(t) = X_0 + \sum_{i \in \mathcal{I}} X_i \text{ with probability } \left(\prod_{i \in \mathcal{I}} p_i\right) \left(\prod_{i \in \mathcal{I}^c} (1 - p_i)\right). \qquad (7.5)$$

Restricting ourselves to the case where each distribution is Gaussian (normal), the observed variable $Y(t)$ is given by

$$Y(t) = \mathcal{N}\left(\mu_0 + \sum_{i \in \mathcal{I}} \mu_i, \; \sigma_0^2 + \sum_{i \in \mathcal{I}} \sigma_i^2\right), \tag{7.6}$$

$$\text{with probability } \left(\prod_{i \in \mathcal{I}} p_i\right)\left(\prod_{i \in \mathcal{I}^c} (1 - p_i)\right).$$

Moreover, it is a straightforward matter to show that

$$E(Y) = \mu_0 + \sum_{i=1}^{k} p_i \mu_i \tag{7.7}$$

$$\text{Var}(Y) = \sigma_0^2 + \sum_{i=1}^{k} p_i \sigma_i^2 + \sum_{i=1}^{k} p_i (1 - p_i)\mu_i^2.$$

A major function of the Shewhart control chart is to find epochs of time which give lots showing characteristics different from those of the "in control" distribution. We note that no assumption is made that this dominant distribution necessarily conforms to any predetermined standards or tolerances. Deming proposes that we find estimates of the dominant μ_0 and σ_0^2 and then find times where lots significantly depart from the dominant. Personal examination of what was unusual about the system when the unusual lot was observed enables us to search for the "assignable cause" of the trouble and fix it. Not a particularly elegant way to proceed perhaps, but plausible *prima facie* and proved amazingly effective by experience.

Consider the flowchart of a production process in Figure 7.2. (For reasons of simplicity, we shall neglect effects of time delays in the flow).

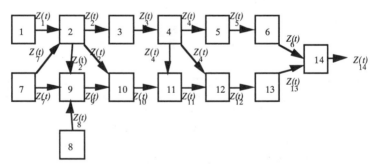

Figure 7.2. Simple Flowchart of Production.

When Deming writes of statistical process control imparting "profound knowledge," he is not resorting to hype or boosterism. On the contrary, this "profound knowledge" is hardnosed and technical. At a very early stage in the optimization process of SPC, we are urged to draw a flowchart of the production process. In many cases, that very basic act (i.e., the composition of the flowchart) is the single most valuable part of the SPC paradigm. Those who are not familiar with real industrial situations might naively assume that a flowchart is composed long before the factory is built and the production begins. Unhappily, such is not the case. To a large extent, the much maligned ISO 9000 protocol for selling goods in the EEC is simply the requirement that a manufacturer write down a flowchart of his or her production process.

The SPC flowchart continually monitors the output of each module and seeks to find atypical outputs at points in time that can be tracked to a particular module. This module is then considered as a candidate for immediate examination for possible suboptimalities, which can then be corrected. The SPC flowchart approach will respond rather quickly to suboptimalities. Let us consider an example.

First of all, let us suppose that in Figure 7.2, the output of module 2 is an aqueous solution where the key measure of Z_2 is the exiting concentration of compound A from module 2. We note that Z_9 is the measured strength of compound B from module 9. In module 10, a compound AB is produced, and the output variable Z_{10} is the strength of that by-product compound. Let us suppose that the end product Z_{14} is the measured reflectivity of a strip of metal. Clearly, the system described above may be one of great complexity. A primitive quality assurance paradigm would simply examine lots of the end product by looking at lot averages of Z_{14} and discarding or reworking lots which do not meet specification standards. If the output is our only measured variable, then any notion of correcting problems upstream of Z_{14} is likely to be attempted by a uniform harangue of the personnel concerned with the management of each of the modules "to do better" without any clue as to how this might be done.

The philosophy of Deming's SPC suggests that we would do much better to find the source of the defect in the system so that it can be rectified. This will be achieved by monitoring each of the intermediate output values $Z_1, Z_2, \ldots, Z_{14}$. Simply looking at Z_{14} will not give us an indication that, say, there is a problem with the control of Z_{10}. This is an example of the truism that "you cannot examine quality into a system."

Let us recall the *Maxim of Pareto:* [1] *The failures in a system are usually the consequence of a few assignable causes rather than the consequence of a general malaise across the system.* Suppose that we make the following further extension of this Maxim: *the failures in a module are usually bunched together in relatively short time epochs, where contamination intervenes,*

---

[1] Interestingly, we may look on Pareto's maxim in the light of Bayes' axiom (postulate 1). If we have a discrete number of causes of failure, Bayes' axiom suggests that we put equal prior probability on each cause. Pareto's maxim (which, although not explicitly stated in his works is fairly deemed to be consistent with them) tells us that it is most likely the probabilities will actually be skewed rather than uniform. The differences represented by the two postulates are consistent with the different philosophies of Bayes and Pareto, the first optimistic and democratic, the second pessimistic and oligarchic.

rather than being uniformly distributed across the time axis. Thus, we are postulating that there will be periods where misfunctioning in a flawed module will be particularly prominent. Statistical process control gives us a simple means for searching for atypical epochs in the record of observations. Whenever we find such atypicality, we will attempt to examine the functioning of the module closely in the hope that we can find the problem and fix it.

Let us return to Figure 7.2. Suppose that we find an atypical epoch of Z_{10}. Since the effect of Z_{10} flows downstream to modules 11 through 14, it may well be that a glitch in Z_{10} will cause glitches in some or all of these as well. However, our best course will be to find the earliest of the modules in a series where the glitch occurs, since that is the one where the assignable cause is most likely to be found (and fixed).

In the example above, let us suppose that we find no atypicality from lot to lot until we get to module 10. Then we also find atypicality in modules 11 through 14. It seems rather clear that we need to examine module 10 for an assignable cause of the system behaving suboptimally. Once the assignable cause is found, it can generally be fixed. Once fixed, it will not soon recur.

7.3   A Compound Test for Higher-Dimensional SPC Data

The basic control chart procedure of Deming is not oriented toward seeing whether a particular lot of items is within predetermined limits, but rather whether the lot is typical of the dominant distribution of items produced. In the one-dimensional case, the interval in which we take a lot sample mean to be "typical," and hence the production process to be "in control" is given by

$$\overline{\overline{x}} - 3\frac{\hat{\sigma}}{\sqrt{n}} \leq \overline{x} \leq \overline{\overline{x}} + 3\frac{\hat{\sigma}}{\sqrt{n}}. \tag{7.8}$$

where n is the lot size, $\overline{x}$ is the mean of a lot, $\hat{\sigma}$ is an estimator for the standard deviation of the dominant population of items, and $\overline{\overline{x}}$ is an estimator for the mean of the dominant population. Assuming that $\overline{x}$ is normally distributed, then the probability that a lot of items coming from the dominant (i.e., "typical") population will fall outside the interval is roughly .002. Generally speaking, because we will usually have plenty of lots, taking the sample variance for each lot, and taking the average of these will give us, essentially, σ^2.

Now let us go from the one-dimensional to the multivariate situation. Following Thompson and Koronacki [9], let us assume that the dominant distribution of output $\mathbf{x}$ data is p-variate normal, that is,

$$f(\mathbf{x}) = |2\pi\boldsymbol{\Sigma}|^{-1/2} \exp\left[-\frac{1}{2}(\mathbf{x} - \boldsymbol{\mu})'\boldsymbol{\Sigma}^{-1}(\mathbf{x} - \boldsymbol{\mu})\right], \qquad (7.9)$$

where $\boldsymbol{\mu}$ is a constant vector and $\boldsymbol{\Sigma}$ is a constant positive definite matrix. By analogy with the use of control charts to find a change in the distribution of the output and/or the input of a module, we can describe a likely scenario of a process going out of control as the mean suddenly changes from, say, $\boldsymbol{\mu}_0$ to some other value. Let us suppose that for jth of N lots, the sample mean is given by $\bar{\mathbf{x}}_j$ and the sample covariance matrix by $\mathbf{S}_j$. Then the natural estimates for $\boldsymbol{\mu}_0$ and $\boldsymbol{\Sigma}$ are

$$\bar{\bar{\mathbf{x}}} = \frac{1}{N}\sum_{j=1}^{N}\bar{\mathbf{x}}_j$$

and

$$\bar{\mathbf{S}} = \frac{1}{N}\sum_{j=1}^{N}\mathbf{S}_j,$$

respectively. Now the Hotelling T^2-like statistic for the jth lot assumes the form

$$T_j^2 = n(\bar{\mathbf{x}}_j - \bar{\bar{\mathbf{x}}})'\bar{\mathbf{S}}^{-1}(\bar{\mathbf{x}}_j - \bar{\bar{\mathbf{x}}}), \qquad (7.10)$$

where $j = 1, 2, \ldots, N$. Alt has shown [1] that

$$\frac{nN - N - p + 1}{p(n-1)(N-1)}T_j^2$$

has the F distribution with p and $nN - N - p + 1$ degrees of freedom. Thus, we consider the jth lot to be out of control if

$$T_j^2 > \frac{p(n-1)(N-1)}{nN - N - p + 1}F_{p, nN - N - p + 1}(\alpha), \qquad (7.11)$$

where $F_{p, nN - N - p + 1}(\alpha)$ is the upper (100α)th percentile of the F distribution with p and $nN - N - p + 1$ degrees of freedom. In SPC, it is generally a fair assumption that N is large, so that we can declare the jth lot to be out of control if

$$T_j^2 > \chi_p^2(\alpha), \qquad (7.12)$$

where $\chi_p^2(\alpha)$ is the upper (100α)th percentile of the χ^2 distribution with p degrees of freedom.

The dispersion matrix (i.e., the covariance matrix of the set of estimates $\hat{\boldsymbol{\mu}}$) is given by

$$\mathbf{V}_{(p \times p)} = \begin{bmatrix} \text{Var}(\hat{\mu}_1) & \text{Cov}(\hat{\mu}_1, \hat{\mu}_2) & \cdots & \text{Cov}(\hat{\mu}_1, \hat{\mu}_p) \\ \text{Cov}(\hat{\mu}_1, \hat{\mu}_2) & \text{Var}(\hat{\mu}_2) & \cdots & \text{Cov}(\hat{\mu}_2, \hat{\mu}_p) \\ \vdots & \vdots & & \vdots \\ \text{Cov}(\hat{\mu}_1, \hat{\mu}_p) & \text{Cov}(\hat{\mu}_2, \hat{\mu}_p) & \cdots & \text{Var}(\hat{\mu}_p) \end{bmatrix}, \qquad (7.13)$$

where,

$$\hat{\mu}_j = \overline{x}_j = \frac{1}{n} \sum_{i=1}^{n} x_{ij} \tag{7.14}$$

for each j.

Let us investigate the power (probability of rejection of the null hypothesis) of the T_j^2 test as a function of the noncentrality:

$$\lambda = (\mu - \mu_0)' \mathbf{V}_{(p \times p)}^{-1} (\mu - \mu_0). \tag{7.15}$$

One can use the approximation [8] for the power of the T_j^2 test:

$$P(\lambda) = \int_{[(p+\lambda)/(p+2\lambda)]\chi_\alpha^2(p)}^{\infty} d\chi^2 \left(p + \frac{\lambda^2}{p + 2\lambda} \right), \tag{7.16}$$

where $d\chi^2(p)$ is the differential of the cumulative distribution function of the central χ^2 distribution with p degrees of freedom and $\chi_\alpha^2(p)$ its $100(1-\alpha)\%$ point. Now, in the current application, any attempt at a numerical approximation technique is unwieldy, due to the fact that we shall be advocating a multivariate strategy based on a battery of nonindependent test. Here is just one of the myriad of instances in real-world applications where simulation can be used to provide quickly and painlessly to the user an excellent estimate of what we need to know at the modest cost of a few minutes crunching on a modern desktop computer.

Current practice is for virtually all testing of p-dimensional data to be carried out by a battery of p one-dimensional tests. Practitioners rightly feel that if glitches are predominant in one or another of the p dimensions, then the information in the p-dimensional multivariate Hotelling statistic will tend to be obscured by the inclusion of channels that are in control.

As an example, let us consider the case where $p = 5$. Thompson and Koronacki [9] have proposed the following compound test.

1. Perform the five one-dimensional tests at nominal Type I error of α = .002 each.

2. Next, perform the 10 two-dimensional tests at nominal $\alpha = .002$ for each.

3. Then perform the 10 three-dimensional tests.

4. Then perform the five four-dimensional tests

5. Finally, perform the one five-dimensional test.

If all the tests were independent, we would expect a pooled Type I error of

$$\alpha = 1 - (1 - .002)^{31} = .06. \tag{7.17}$$

Table 7.1. Type I Errors of Compound Test								
p	1-d Tests	$n = 5$	$n = 10$	$n = 15$	$n = 20$	$n = 50$	$n = 100$	$n = 200$
2	.004	.00434	.00466	.00494	.00508	.00512	.00528	.00546
3	.006	.00756	.00722	.00720	.00720	.00704	.00826	.00802
4	.008	.01126	.01072	.01098	.01098	.01108	.01136	.01050
5	.010	.01536	.01582	.01552	.01544	.01706	.01670	.01728

However, the tests are not really independent [so we cannot use (7.16)]. For dimensions two through five, using uncorrelated vector components, we show in Table 7.1, the Type I errors based on 50,000 simulations per table entry. In the second column, we show the Type I errors if only the one-dimensional tests are carried out (i.e., $p \times .002$). The subsequent columns give the Type I errors for various lot sizes (n) assuming we use all the possible $2^p - 1$ tests. The resulting Type I errors for the pooled tests for all dimensions are not much higher than those obtained simply by using only the one-dimensional tests. We note that in the five-dimensional case, if we use only the five one-dimensional tests, we have a Type I error of .01. Adding in all the two-dimensional, three-dimensional, four-dimensional and five-dimensional tests does not even double the Type I error. As a practical matter, a user can, if desired, simply continue to use the one-dimensional tests for action, reserving the compound higher-dimensional tests for exploratory purposes.

We note here how the use of simulation, essentially as a "desk accessory," enabled us quickly to determine the downside risks of using a new testing strategy. Slogging through analytical evaluations of the compound test would have been a formidable task indeed. Using a simulation-based evaluation, we were able quickly to see that the price for the compound test was small enough that it should be seriously considered.

7.4 Rank Testing with Higher-Dimensional SPC Data

In statistical process control, we are looking for a difference in the distribution of a new lot, anything out of the ordinary. That might seem to indicate a nonparametric density estimation—based procedure. But the general ability to look at averages in statistical process control indicates that for many situations, the central limit theorem enables us to use procedures that point to distributions somewhat close to the normal distribution as the standards. In the case where data is truly normal, the functional form of the underlying density can be based exclusively on the mean vector and the covariance matrix. However, as we show below, it is a rather easy matter to create multivariate tests that perform well in the normal case and in heavy tailed departures from normality.

Consider the case where we have a base sample of N lots, each of size n, with the dimensionality of the data being given by p. For each of these lots, compute the sample mean $\overline{\mathbf{X}}_i$ and sample covariance matrix $\mathbf{S}_i$. Moving on, compute the average of these N sample means, $\overline{\overline{\mathbf{X}}}$, and the average of the sample covariance matrices $\overline{\mathbf{S}}$. Then, use the transformation

$$\mathbf{Z} = \overline{\mathbf{S}}^{-1/2}(\mathbf{X} - \overline{\overline{\mathbf{X}}}), \tag{7.18}$$

which transforms $\mathbf{X}$ into a variate with approximate mean $\mathbf{0}$ and and approximate covariance matrix $\mathbf{I}$. Next, apply this transformation to each of the N lot means in the base sample. For each of the transformed lots, compute the transformed mean and covariance matrix, $\overline{\mathbf{Z}}_i$ and $\mathbf{S}_{\mathbf{Z}_i}$, respectively. For each of these, apply, respectively, the location norm

$$\|\overline{\mathbf{Z}}_i\|^2 = \sum_{j=1}^{p} \overline{Z}_{i,j}^{\;2}, \tag{7.19}$$

and scale norm

$$\|\mathbf{S}_i\|^2 = \sum_{j=1}^{p}\sum_{l=1}^{p} S_{i,j,l}^{\;2}. \tag{7.20}$$

Now, if a new lot has location norm higher than any of those in the base sample, we flag it as atypical. If its scale norm is greater than those of any lot in the base sample, we flag it as atypical. The Type I error of either test is given, approximately, by $1/(N+1)$, that of the combined test is given very closely by

$$1 - \left(1 - \frac{1}{N+1}\right)^2 = \frac{2N+1}{(N+1)^2}. \tag{7.21}$$

Let us now compare the performance of the location rank test with that of the parametric likelihood ratio test when we have as the generator of the 'in control" data a p-variate normal distribution with mean $\mathbf{0}$ and covariance matrix $\mathbf{I}$, the identity. We consider as alternatives "slipped" normal distributions, each with covariance matrix $\mathbf{I}$ but with a translated mean each of whose components is equal to the "slippage" μ. In Figure 7.3, using 20,000 simulations of 50 lots of size 5 per slippage value to obtain the base information, we compute the efficiency of the rank test to detect a shifted 51st lot relative to that of the likelihood ratio test [i.e., the ratio of the power of the rank test to that of the $\chi^2(p)$ test (where p is the dimensionality of the data set)]. In other words, here, we assume that both the base data and the lots to be tested have identity covariance matrix and that this matrix is known. We note that the efficiency of the rank test here, in a situation favorable to the likelihood ratio test, is close to 1, with generally improving performance as the dimensionality increases. Here, we have used the critical values from tables of the χ^2 distribution. For such a situation, the χ^2 is the likelihood ratio test, so in a sense this is a very favorable case for the parametric test. In Figure 7.3 we apply the location test only for the data simulation delineated above.

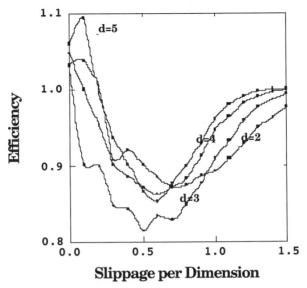

Figure 7.3. Monte Carlo Estimates of Efficiencies (Normal Data).

Next, we consider applying the rank test for location to $\mathbf{t}(3)$ data generated in the obvious manner as shown in Figure 7.4. First, we generate a χ^2 variable v with 3 degrees of freedom. Then we generate p independent univariate normal variates $\mathbf{X}' = (X_1, X_2, \ldots, X_p)$ from a normal distribution with mean 0 and variance 1. If we wish to have a mean vector $\boldsymbol{\mu}$ and covariance matrix $\mathbf{I}$,

$$\mathbf{t} = \frac{\mathbf{X}}{\sqrt{v/3}} + \boldsymbol{\mu} \tag{7.22}$$

will have a shifted $\mathbf{t}$ distribution with 3 degrees of freedom.

Once again the rank test performs well when its power is compared to that of the parametric test *even though we have computed the critical value for the parametric test assuming the data is known to be* $\mathbf{t}(3)$. We should remember, however, that if we had assumed (incorrectly) that the data were multivariate normal, the likelihood ratio test would have been quite different and its results very bad. (Naturally, as the lot size becomes quite large, the central limit theorem will render the normal theory-based test satisfactory.) The rank test performs well whether the data are normal or much more diffuse, and it requires no prior information as to whether the data is normal or otherwise.

So far, we have been assuming that both the base lots and the new lots were known to have identity covariance matrices. In such a case, the appropriate parametric test is χ^2 if the data is normal, and if it is not, we have employed simulation techniques to find appropriate critical values for the distribution in question. Now, however, we shift to the situation where we believe that the covariance matrices of the new lots to be sampled may not be diagonal. We have been assuming that the base lots (each of size 5) are drawn from $\mathcal{N}(\mathbf{0}, \mathbf{I})$. The sampled (bad) lot is drawn from $\mathcal{N}(\boldsymbol{\mu}, \boldsymbol{\Sigma})$, where

$$\boldsymbol{\mu} = \begin{pmatrix} \mu \\ \mu \\ \vdots \\ \mu \end{pmatrix} \tag{7.23}$$

and

$$\boldsymbol{\Sigma} = \begin{pmatrix} 1 & .8 & .8 & \cdots & .8 \\ .8 & 1 & .8 & \cdots & .8 \\ .8 & .8 & 1 & \cdots & .8 \\ \vdots & \vdots & \vdots & \vdots & \vdots \\ .8 & .8 & .8 & \cdots & \vdots \end{pmatrix}. \tag{7.24}$$

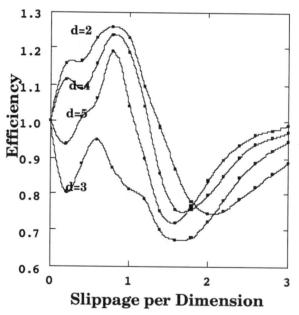

Figure 7.4. Monte Carlo Estimates of Efficiencies (t(3) Data).

Thus, we are considering the case where the lot comes from a multivariate normal distribution with equal slippage in each dimension and a covariance matrix that has unity marginal variances and covariances .8. In Figure 7.5, we note the relative power of the "location" rank test when compared to that of the Hotelling T^2 procedure. The very favorable performance of the rank test is largely due to the effect that it picks up not only changes in location but also departures in the covariance matrix of the new lot from that of the base lots. The basic setting of statistical process control lends itself very naturally to the utilization of normal distribution theory, since computation of lot averages is so customary. But as we have seen, for modest lot sizes it is possible to run into difficulty if the underlying distributions have heavy tails.

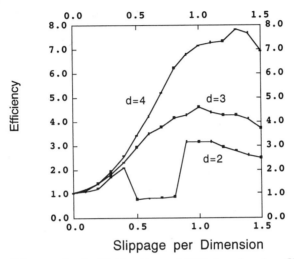

Figure 7.5. Monte Carlo Estimates of Efficiencies for Correlated Data.

In the construction of these rank tests by Thompson, Lawera and Koronacki, [7],[9]. a substantial number of unsuccessful tests was examined before noting the testing procedure explicated here. Again, the utility of simulation is demonstrated. It is all very well to try an analytical approach for such tests, examining, for example, asymptotic properties. But few of us would willingly expend months of effort on tests which might well "come a cropper." Simulation gives us a means of quickly stressing potentially useful tests quickly and efficiently.

7.5   A Robust Estimation Procedure for Location in Higher Dimensions

Let us recall that, in the philosophy of Deming, one should not waste much time in determining whether a lot conforms to some predetermined standards. Many have thought themselves inspired because they came up with very strict standards for, say, manufacturing automobile transmissions. The very statement of strenuous goals is thought by many contemporary American managers (not to mention directors of Soviet *five-year plans* in a bygone age, or presidents of American universities in this age) to be a constructive act. SPC does not work that way. In SPC we seek to see epochs when lots appear to have been produced by some variant of the dominant in-control process which produces most of the lots.

But, how shall we attempt to look into the wilderness of the past record of lots and determine what actually is, say, the location of the dominant output? It is not an easy task. Obviously, to the extent that we include out-of-control lots in our estimate of the location of the in-control lots we will have contaminated the estimate.

The following King of the Mountain algorithm of Lawera and Thompson [7] (see also Thompson and Koronacki, [9]) appears to be promising:

"King of the Mountain" Trimmed Mean Algorithm

Set the counter M equal to the historical proportion of bad lots times number of lots.

For N lots compute the vector sample means of each lot $\{\overline{\mathbf{X}}_i\}_{i=1}^N$.

1. Compute the pooled mean of the means $\overline{\overline{\mathbf{X}}}$.

Find the two sample means farthest apart in the cloud of lot means.

From these two sample means, discard the farthest from $\overline{\overline{\mathbf{X}}}$.

Let $M = M - 1$ and $N = N - 1$.

If the counter is still positive, go to 1; otherwise exit and print out $\overline{\overline{\mathbf{X}}}$ as $\overline{\overline{\mathbf{X}}}_T$.

To examine the algorithm, we examine a mixture distribution of lot means

$$\gamma \mathcal{N}(\mathbf{0}, \mathbf{I}) + (1 - \gamma)\mathcal{N}(\boldsymbol{\mu}, \mathbf{I}). \tag{7.25}$$

Here we assume equal slippage in each dimension, that is,

$$(\boldsymbol{\mu}) = (\mu, \mu, \dots, \mu). \tag{7.26}$$

Let us compare the trimmed mean procedure $\overline{\overline{X}}_T$ with the customary procedure of using the untrimmed mean $\overline{\overline{X}}$. In Tables 7.2 and 7.3 we show for dimensions two, three, four, and five, the average MSEs of the two estimators when $\gamma = .70$ for simulations of size 1000.

Table 7.2. MSEs for 50 Lots: $\gamma = .7$							
	d=2	d=2	d=3	d=3	d=4	d=4	d=5	d=5
μ	$\overline{\overline{X}}_T$	$\overline{\overline{X}}$	$\overline{\overline{X}}_T$	$\overline{\overline{X}}$	$\overline{\overline{X}}_T$	$\overline{\overline{X}}$	$\overline{\overline{X}}_T$	$\overline{\overline{X}}$
1	0.40	0.94	0.43	1.17	0.46	1.42	0.54	1.72
2	0.21	1.24	0.17	1.62	0.17	2.00	0.18	2.37
3	0.07	1.85	0.09	2.67	0.12	3.49	0.15	4.33
4	0.06	3.01	0.09	4.48	0.12	5.93	0.14	7.41
5	0.05	4.61	0.09	6.88	0.11	9.16	0.15	11.52
6	0.06	6.53	0.08	9.85	0.12	13.19	0.14	16.50
7	0.06	8.94	0.09	13.41	0.11	17.93	0.14	22.33
8	0.06	11.58	0.08	17.46	0.11	23.27	0.14	28.99
9	0.06	14.71	0.08	22.03	0.11	29.40	0.15	36.67
10	0.06	18.13	0.08	27.24	0.11	36.07	0.15	45.26

Table 7.3. MSEs for 100 Lots: $\gamma = .7$							
	d=2	d=2	d=3	d=3	d=4	d=4	d=5	d=5
μ	$\overline{\overline{X}}_T$	$\overline{\overline{X}}$	$\overline{\overline{X}}_T$	$\overline{\overline{X}}$	$\overline{\overline{X}}_T$	$\overline{\overline{X}}$	$\overline{\overline{X}}_T$	$\overline{\overline{X}}$
1	0.28	0.71	0.30	0.94	0.32	1.16	0.34	1.34
2	0.05	0.88	0.06	1.30	0.07	1.74	0.08	2.10
3	0.03	1.72	0.05	2.58	0.06	3.37	0.07	4.25
4	0.03	2.95	0.04	4.41	0.06	5.86	0.07	7.34
5	0.03	4.56	0.04	8.25	0.06	9.13	0.08	11.39
6	0.03	6.56	0.04	8.31	0.06	13.08	0.07	16.35
7	0.03	8.87	0.04	13.28	0.06	17.67	0.07	22.16
8	0.03	11.61	0.04	17.32	0.06	23.17	0.07	28.86
9	0.03	14.67	0.04	21.98	0.06	29.28	0.07	36.61
10	0.03	18.04	0.04	27.12	0.06	36.06	0.07	45.13

In Tables 7.4 and 7.5 we show the MSEs of the trimmed mean and the customary pooled sample mean for the case where $\gamma = .95$.

Table 7.4. MSEs for 50 Lots: $\gamma = .95$							
	d=2	d=2	d=3	d=3	d=4	d=4	d=5	d=5
μ	$\overline{\overline{X}}_T$	$\overline{\overline{X}}$	$\overline{\overline{X}}_T$	$\overline{\overline{X}}$	$\overline{\overline{X}}_T$	$\overline{\overline{X}}$	$\overline{\overline{X}}_T$	$\overline{\overline{X}}$
1	0.05	0.10	0.07	0.15	0.09	0.18	0.12	0.24
2	0.05	0.11	0.07	0.16	0.09	0.20	0.11	0.26
3	0.04	0.11	0.06	0.17	0.08	0.22	0.11	0.28
4	0.04	0.13	0.06	0.19	0.08	0.26	0.11	0.33
5	0.04	0.16	0.06	0.24	0.08	0.32	0.11	0.41
6	0.04	0.19	0.06	0.29	0.09	0.40	0.10	0.49
7	0.04	0.24	0.06	0.35	0.08	0.48	0.11	0.61
8	0.04	0.29	0.06	0.42	0.08	0.58	0.10	0.71
9	0.04	0.33	0.06	0.51	0.08	0.68	0.11	0.86
10	0.042	0.39	0.06	0.59	0.08	0.79	0.11	1.01

Table 7.5. MSEs for 100 Lots: $\gamma = .95$							
	d=2	d=2	d=3	d=3	d=4	d=4	d=5	d=5
μ	$\overline{\overline{X}}_T$	$\overline{\overline{X}}$	$\overline{\overline{X}}_T$	$\overline{\overline{X}}$	$\overline{\overline{X}}_T$	$\overline{\overline{X}}$	$\overline{\overline{X}}_T$	$\overline{\overline{X}}$
1	0.02	0.05	0.04	0.09	0.05	0.12	0.06	0.14
2	0.02	0.07	0.03	0.10	0.05	0.14	0.05	0.16
3	0.02	0.09	0.03	0.14	0.04	0.18	0.05	0.23
4	0.02	0.13	0.03	0.20	0.05	0.27	0.06	0.34
5	0.02	0.18	0.03	0.27	0.04	0.37	0.06	0.46
6	0.02	0.24	0.03	0.38	0.04	0.49	0.05	0.60
7	0.02	0.31	0.03	0.47	0.04	0.62	0.06	0.80
8	0.02	0.40	0.03	0.60	0.04	0.80	0.05	1.00
9	0.02	0.50	0.03	0.74	0.04	1.00	0.05	1.21
10	0.02	0.60	0.03	0.90	0.05	1.20	0.05	1.50

If the level of contamination is substantial (e.g., if $1 - \gamma = .3$), the use of a trimming procedure to find a base estimate of the center of the in-control distribution contaminated by observations from other distributions may be strongly indicated. For more modest but still significant levels of contamination (e.g., if $1-\gamma=.05$), simply using $\overline{\overline{X}}$ may be satisfactory. We note that the trimming procedure considered here is computer intensive and is not realistic to be performed on the usual hand-held scientific calculator. However, it is easily computed on a personal computer or workstation. Since the standards for rejecting the null hypothesis that a lot is in control are generally done by off-line analysis on a daily basis, we do not feel that the increase in computational complexity should pose much of a logistical problem.

Problems

7.1. It is desired to simulate a contamination model for training purposes. We wish to produce sheets of aluminum with thickness 1 mm. Suppose the dominant distribution is given by $\mathcal{N}(1,1)$. Time is divided up into epochs of 10 minutes. Contamination will occur in each epoch with probability γ. The contaminating distribution will be $\mathcal{N}(\mu,\sigma^2)$. Simulate data for a variety of γ's, μ's, and σ's. Using Shewhart control charting (7.8), see how effective you are in finding lots that are from contamination periods.

7.2. Let us consider the case where there are four measurables in a production process. After some transformation, the in control situation would be represented by a normal distribution with mean

$$\mu = \begin{pmatrix} \mu_1 & = & 0 \\ \mu_2 & = & 0 \\ \mu_3 & = & 0 \\ \mu_4 & = & 0 \end{pmatrix}$$

and covariance matrix

$$\Sigma = \mathbf{I} = \begin{pmatrix} \sigma_{11}=1 & \sigma_{12}=0 & \sigma_{13}=0 & \sigma_{14}=0 \\ \sigma_{12}=0 & \sigma_{22}=1 & \sigma_{23}=0 & \sigma_{24}=0 \\ \sigma_{13}=0 & \sigma_{23}=0 & \sigma_{33}=1 & \sigma_{34}=0 \\ \sigma_{14}=0 & \sigma_{24}=0 & \sigma_{24}=0 & \sigma_{44}=1 \end{pmatrix}.$$

We wish to examine the effectiveness of the battery of tests procedure in Section 7.3, for the following situations, each with lot sizes of 10:
(a) There is 5% contamination from the Gaussian distribution with

$$\mu = \begin{pmatrix} \mu_1 & = & 1 \\ \mu_2 & = & 1 \\ \mu_3 & = & 1 \\ \mu_4 & = & 1 \end{pmatrix}$$

and covariance matrix

$$\Sigma = \mathbf{I}.$$

(b) There is 5% contamination from a Gaussian distribution with

$$\mu = \begin{pmatrix} \mu_1 & = & 0 \\ \mu_2 & = & 2 \\ \mu_3 & = & -2 \\ \mu_4 & = & 0 \end{pmatrix}$$

and covariance matrix

$$\Sigma = I.$$

(c) There is 10% contamination from a Gaussian distribution with

$$\mu = \begin{pmatrix} \mu_1 & = & 1 \\ \mu_2 & = & 1 \\ \mu_3 & = & 1 \\ \mu_4 & = & 1 \end{pmatrix}$$

and covariance matrix

$$\Sigma = \begin{pmatrix} \sigma_{11} = 1 & \sigma_{12} = .6 & \sigma_{13} = .6 & \sigma_{14} = .6 \\ \sigma_{12} = .6 & \sigma_{22} = 1 & \sigma_{23} = .6 & \sigma_{24} = .6 \\ \sigma_{13} = .6 & \sigma_{23} = .6 & \sigma_{33} = 1 & \sigma_{34} = .6 \\ \sigma_{14} = 0 & \sigma_{24} = 0 & \sigma_{24} = 0 & \sigma_{44} = 1 \end{pmatrix}.$$

7.3. Next, let us consider the case where there are three measurables in a production process. After some transformation, the in control situation would be represented by a normal distribution with mean

$$\mu = \begin{pmatrix} 0 \\ 0 \\ 0 \end{pmatrix}$$

and covariance matrix

$$\Sigma = I.$$

There is 10% contamination of lots (i.e., 10% of the lots are from the contaminating distribution). The contaminating distribution is multivariate t with 3 degrees of freedom (see (7.22)) and translated to

$$\mu = \begin{pmatrix} \mu_1 & = & 1 \\ \mu_2 & = & 1 \\ \mu_3 & = & 1 \end{pmatrix}.$$

If lot sizes are 8, compare the effectiveness of the battery of tests above with the rank test given in Section 7.4.

7.4. You have a six-dimensional data set which turns out to be

$$\gamma \mathcal{N}(0, I) + (1 - \gamma) t_3(1, I).$$

The problem is to recover the mean of the uncontaminated Gaussian distribution for the case where $\gamma = .2$ Lot sizes are 10. See how well the King of the Mountain Algorithm works for this case compared to using pooled lot means.

References

[1] Alt, F.B. (1985). "Multivariate quality control," in *Encyclopedia of Statistical Sciences*, **6**, S. Kotz, and N. Johnson eds. New York: John Wiley & Sons, 110–122,

[2] Banks, D. (1993). "Is industrial statistics out of control?" *Statist. Sc.*, **8**, 356–409.

[3] Deming, W. E. (1982). *Quality, Productivity and Competitive Position.* Cambridge, Mass.: Center for Advanced Engineering Study, Massachusetts Institute of Technology.

[4] Deming, W.E. (1986). *Out of the Crisis.* Cambridge, Mass.: Center for Advanced Engineering Study, Massachusetts Institute of Technology.

[5] Juran, J.M. (1995) "The Non-Pareto Principle; Mea Culpa," *Quality Prog.*, 149–150.

[6] Koronacki, J. and Thompson, J.R. (1997). "Statystyczne sterowanie procesem wielowymiarowym" ("Statistical process control in high dimensions"), *Problemy Jakości*, **xxix**, 8–15.

[7] Lawera, M. and Thompson, J. R. (1993). "Some problems of estimation and testing in multivariate statistical process control," *Proceedings of the Thirty-Fourth Conference on the Design of Experiments in Army Research Development and Testing.* Research Triangle Park, N.C.: Army Research Office, 99–126.

[8] Stuart, A. and Ord, J.K. (1991). *Kendall's Advanced Theory of Statistics,* **2**. New York: Oxford University Press, 870.

[9] Thompson, J.R. and Koronacki, J. (1992). *Statistical Process Control for Quality Improvement.* New York: Chapman & Hall.

[10] Thompson, J R. (1989). *Empirical Model Building.* New York: John Wiley & Sons.

Chapter 8

Noise and Chaos

> Anything could be true. The so-called laws of nature were nonsense. The law of gravity was nonsense. "If I wished," O'Brien had said, "I could float off this floor like a soap bubble." Winston worked it out. "If he *thinks* he floats off the floor, and I simultaneously *think* I see him do it, then the thing happens."
>
> George Orwell, *1984*

8.1 Introduction

Postmodernism is one of the latest intellectual schools to be inflicted on the West since the French Revolution. Postmodernism does more than cast doubt on objective reality; it flatly denies it. The perception of a particular reality is deemed to be highly subjective and hugely nonstationary, even for the same observer. Those schooled in the implicit Aristotelian modality of reason and logic (and that still includes most scientists) can easily follow the postmodernist train of thought to the point where they note that the assumptions of the postmodernists really bring one to such a level of chaos that conversation itself becomes impractical and useless. Fifteen years ago one could find few people presenting papers on such subjects as *postmodern science*; this has become more frequent lately. Postmodernism, as the latest of the pre-Socratic assaults on reason based on facts, was gently smiled at by scientists simply as a bizarre attempt of their humanistic colleagues to appear to be doing something useful. In the last few years, however, the leakage of postmodern modalities of thinking into the sciences has increased significantly. When one looks in library offerings under such topics as chaos theory or fuzzy logic, one frequently finds "postmodern" as a correlative listing. For some of us Aristotelians, this tendency is alarming. But many mathematicians are pleased to find that their abstract ideas are appreciated by literati and the popular press.

Every 20 years or so, there crops up a notion in *mainstream science* which is every bit as antilogical and antirealistic as postmodernism itself. These notions either do not admit of scientific validation or, for some reason, are somehow exempted from it. Once these *new ideas* have been imprimaturized by the scientific establishment and developed into a systematology, large amounts of funding are provided by government agencies, and persons practicing such arts are handsomely rewarded by honors and promotion until lack of utility causes the systematology to be superseded or subsumed by some other *new idea*.

It is intriguing (and should be noncomforting to stochasticians) that such *new ideas* tend to have as a common tendency the promise that practitioners of the new art will be able to dispense with such primitive notions as probability. Stochastics, after all, was simply a *patch*, an empirical artifice, to get around certain bumps in the scientific road—which jumps have now been smoothed level by the new art.

Frequently, not only will the new art not give any hope for solving a scientific problem, it will actually give comfort to an economist, say, or a meteorologist, or the members of some area of science which has not lived up to hopes, over the *fact* that their area simply does not lend itself to the solution of many of its most fundamental problems. It does not merely say that these problems are hard, but it argues that they cannot be solved now or ever. Those who have labored for decades on these hard problems may now safely down shovels, knowing that they gave their best to do what could not be done.

We have seen a number of scientists take problems that were supposed to be insoluble and solve them. The "hopeless" hairtrigger of nuclear war was obviated by one of the founders of Monte Carlo techniques, Herman Kahn. Kahn essentially created the escalation ladder, giving the great powers a new grammar of discourse which enabled staged response to crises.

High-speed computing broke out of "natural" bounds of feasibility by the invention first of the transistor and then the microchip. Weather forecasting is still admittedly primitive, but apparently we are now in a position to carry out such tasks as forecasting severe as opposed to mild hurricane seasons.

To introduce the subject of chaos, let us first consider the Mandelbrot model,

$$z_{n+1} = z_n^2 + c, \qquad (8.1)$$

where z is a complex variable and c is a complex constant. Let us start with

$$z_1 = c = -.339799 + .358623i.$$

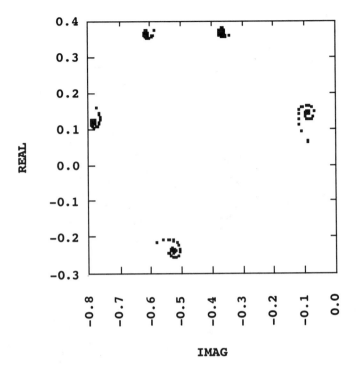

Figure 8.1. Mendelbrot Set.

The iterative structure of this set is given in Figure 8.1. All the points exist in thin curved manifolds. Most of the complex plane is empty. And, to make things more interesting, we do not trace smoothly within each manifold, but rather jump from manifold to manifold. It is as though we started to drive to the opera hall in Houston, but suddenly found ourselves in downtown Vladivostok. Of course, it could be interesting for somebody who fantasizes about zipping from one part of the galaxy to another.

Now, the Mandelbrot model may itself be questioned as a bit bizarre when describing any natural phenomenon. Were it restricted to the reals, then it would be, for a growth model, something that we would be unlikely to experience, since it is explosive. But let that pass. Since we are in the realm of *Gedankenspiel* anyway, let us see what happens when we introduce noise.

Consider the model

$$z_{n+1} = z_n^2 + c + .2u_{n+1} + .2v_{n+1}, \tag{8.2}$$

where the u and v are independent $\mathcal{N}(0,1)$ variables. We have been able to penetrate throughout the former empty space, and now we do not make quantum leaps from Houston to Vladivostok. Here, in the case of a formal mathematical structure, we simply note in passing how the introduction of noise can remove apparent pathology. Noise can and does act as a powerful smoother in many situations based on aggregation (Figure 8.2).

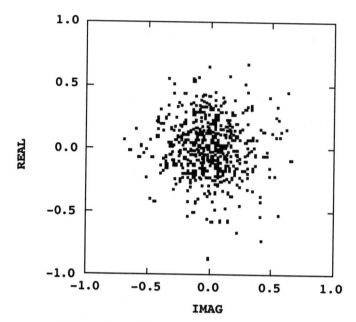

Figure 8.2. Mendelbrot Plus Noise.

8.2   The Discrete Logistic Model

Let us turn now to an example of a model of a real-world phenomenon, namely the growth of a population with finite food supply. One of the earliest was the 1844 logistic model of Verhulst:

$$\frac{dX}{dt} = X(\alpha - X),\tag{8.3}$$

where α essentially represents the limit, in population units, of the food supply. A solution to this model was obtained by Verhulst and is simply

$$X(t) = \frac{\alpha X(0)\exp(\alpha t)}{\alpha + X(0)[\exp(\alpha t) - 1]}.\tag{8.4}$$

Naturally, this model is only an approximation to the real growth of a population, but the mathematical solution is perfectly regular and without pathology.

Lorenz [2] has examined a discrete version of the logistic model:

$$X_n = X_{n-1}(a - X_{n-1}) = -X_{n-1}^2 + aX_{n-1}.\tag{8.5}$$

Using $X_0 = a/2$, he considers the time average of the modeled population size:

$$\overline{X} = \lim_{N \to \infty} \frac{1}{N} \sum_{n=1}^{N} X_n. \tag{8.6}$$

For a values below $1 + \sqrt{6}$, the graph of $\overline{X}$ behaves quite predictably. Above this value, great instability appears, as we show in Figure 8.3.

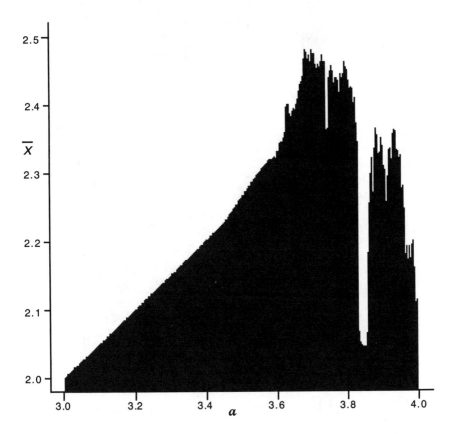

Figure 8.3. Discrete Logistic Model.

We note in Figures 8.4 and 8.5 how this *fractal* structure is maintained at any level of microscopic examination we might choose.

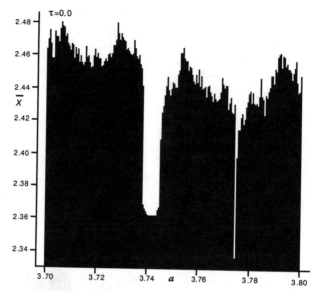

Figure 8.4. Discrete Logistic Model at Small Scale.

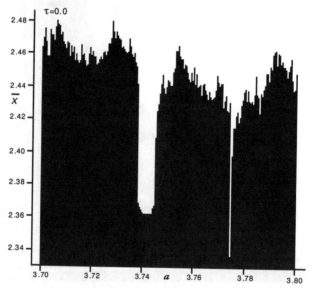

Figure 8.5. Discrete Logistic Model at Very Small Scale.

Let us look at Figure 8.3 in the light of real-world ecosystems. Do we know of systems where we increase the food supply slightly and the supported population crashes, we increase it again and it soars? What should be our point of view concerning a model that produces results greatly at variance with the real world? And we recall that actually we have a perfectly good continuous 150 year old solution to the logistic equation. The use of the

discrete logistic model is really natural only in the sense that we wish to
come up with a model that can be put on a digital computer. In the case
of chaos theory it is frequently the practice of enthusiasts to question not
the model but the reality. So it is argued that, in fact, it is the discrete
model which is the more natural. Let us walk for a time in that country.

For the kinds of systems the logistic model was supposed to describe, we
could axiomatize by a birth-and-death process as follows:

$$\begin{aligned}
P(\text{birth in } [t, t + \Delta t)) &= \beta(\gamma - X)X\Delta t \\
P(\text{death in } [t, t + \Delta t)) &= \eta X \Delta t.
\end{aligned} \qquad (8.7)$$

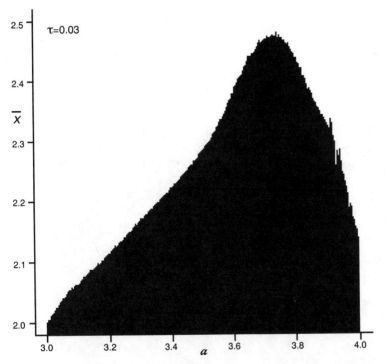

Figure 8.6. Discrete Logistic Model Plus Noise.

Perhaps Verhulst would have agreed that what he had in mind to do was
to aggregate from such microaxioms but had not the computational ability
to do so. Equation (8.5) was the best he could do. We have the computing
power to deal directly with the birth-and-death model. However, we can
make essentially the same point by adding a noise component to the logistic
model. We do so as follows:

$$\begin{aligned}
X_n &= X_{n-1}(a - X_{n-1}) + \mu_{n-1}X_{m-1} \\
&= X_{n-1}([a + \mu_{n-1}] - X_{n-1}).
\end{aligned} \qquad (8.8)$$

where μ_{n-1} is a random variable from the uniform distribution on $(-\tau, \tau)$.

As a convenience, we add a bounceback effect for values of the population less than zero. Namely, if the model drops the population to $-\epsilon$, we record it as $+\epsilon$. In Figure 8.6 we note that the stochastic model produces no chaos (the slight fuzziness is due to the fact that we have averaged only 5000 successive X_n values). Nor is there fractal structure at the microscopic level, as we show in Figure 8.7 (using 70,000 successive X_n values).

Clearly, the noisy version of (8.5) is closer to the real world than the purely deterministic model. The food supply changes; the reproductive rate changes; the population is subject to constant change. However, the change itself induces stability into the system. The extreme sensitivity to a in Figure 8.3 is referred to as the butterfly effect. The notion is that if (8.5) described the weather of the United States, then one butterfly flying across a backyard could dramatically change the climate of the nation. Such an effect, patently absurd, is a failure of the model (8.5), not of the real world.

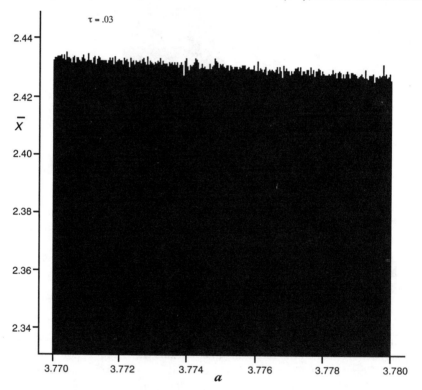

Figure 8.7. Discrete Logistic Model Plus Noise at Small Scale.

8.3   A Chaotic Convection Model

In 1963, after rounding off initial conditions, Lorenz [1] discovered that the following model was extremely sensitive to the initial value of (x, y, z).

$$\frac{dx}{dt} = 10(y - x)$$

$$\frac{dy}{dt} = -xz + 28x - y \qquad\qquad (8.9)$$

$$\frac{dz}{dt} = xy - \frac{8}{3}z.$$

In Figure 8.8 we show a plot of the system for 2000 steps using $\Delta t = .01$ and $\tau = 0.0$.

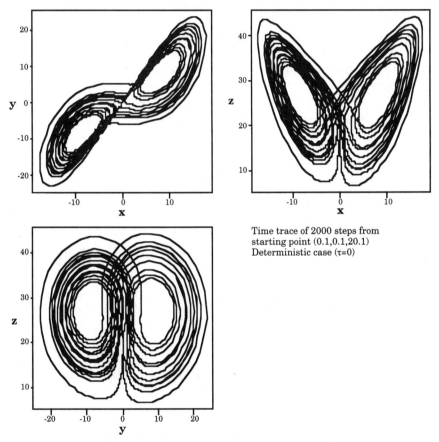

Time trace of 2000 steps from starting point (0.1,0.1,20.1) Deterministic case (τ=0)

Figure 8.8. Lorenz Weather Model.

We observe the nonrepeating spiral which characterizes this chaotic model. The point to be made here is that, depending on where one starts the process, the position in the remote future will be very different. Lorenz uses this model to explain the poor results one has in predicting the weather. Many have conjectured that such a model corresponds to a kind of uncertainty principle operating in fields ranging from meteorology to economics. Thus, it is opined that in such fields, although a deterministic model may accurately describe the system, there is no possibility of making meaningful long range forecasts, since the smallest change in the initial conditions dramatically changes the end result. The notion of such an uncertainty principle has brought comfort to such people as weather scientists and econometricians who are renowned for their inability to make useful forecasts. What better excuse for poor performance than a mathematical basis for its inevitability?

The philosophical implications of (8.9) are truly significant. Using (8.9), we know precisely what the value of (x, y, z) will be at any future time if we know precisely what the initial values of these variables are (an evident impossibility). We also know that the slightest change in these initial values will dramatically alter the result in the remote future. Furthermore, (8.9) essentially leads us to a dead end. If we believe this model, then it cannot be improved, for if at some time a better model for forecasting were available so that we really could know what (x, y, z) would be at a future time, then, since the chaos spirals are nonrepeating, we would be able to use our knowledge of the future to obtain a precise value of the present. Since infinite precision is not even a measurement possibility, we arrive at a practical absurdity.

If one accepts the ubiquity of chaos in the real world (experience notwithstanding), then one is essentially driven back to pre-Socratic modalities of thought, where experiments were not carried out, since it was thought that they would not give reproducible results. Experience teaches us that, with few exceptions, the models we use to describe reality are only rough approximations. Whenever a model is thought to describe a process completely, we tend to discover, in retrospect, that factors were missing, that perturbations and shocks entered the picture which had not been included in the model. A common means of trying to account for such phenomena is to assume that random shocks of varying amplitudes are constantly being applied to the system. Let us, accordingly, consider a discretized noisy version of (8.9):

$$
\begin{aligned}
x_n &= (1 + \mu_{x,n-1})x_{n-1} + \Delta t\, 10(y_{n-1} - x_{n-1}) \\
y_n &= (1 + \mu_{y,n-1})y_{n-1} + \Delta t(-x_{n-1}y_{n-1} + 28x_{n-1} - y_{n-1}) \quad (8.10) \\
z_n &= (1 + \mu_{z,n-1})z_{n-1} + \Delta t\left(x_{n-1}y_{n-1} - \frac{8}{3}z_{n-1}\right),
\end{aligned}
$$

where the μ's are independently drawn from a uniform distribution on the interval from $(-\tau, \tau)$.

We consider in Figure 8.9 the final point at the 2500th step of each of 1000 random walks using the two initial points $(0.1, 0.1, 20.1)$ and $(-13.839, -6.66, 40.289)$, with $\tau = 0.0001$. These two starting points are selected since, in the deterministic case, the end results are dramatically different. In Figure 8.10, we show quantile-quantile plots for the two cases. We note that, just as we would expect in a model of the real world, the importance of the initial conditions diminishes with time, until, as we see from Figures 8.9 and 8.10, the distribution of end results is essentially independent of the initial conditions.

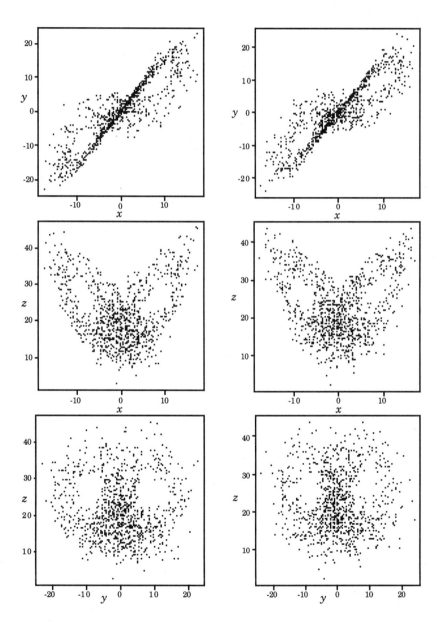

Figure 8.9. 10,000th Step for Each of 1000 Time Traces.

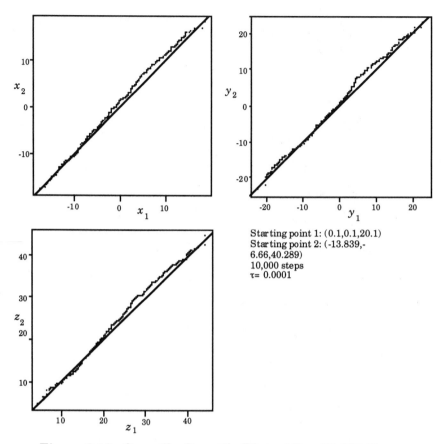

Starting point 1: (0.1,0.1,20.1)
Starting point 2: (-13.839,-6.66,40.289)
10,000 steps
$\tau = 0.0001$

Figure 8.10. Quantile-Quantile Plots After 10,000 Steps.

8.4 Conclusions

It is intriguing that so many scientists have been drawn to a theory which promises such a chaotic world that, if it were the real one, one would not even be able to have a conversation on the merits of the notion. Nevertheless, chaos theory should cause us to rethink aggregation to a supposed mean trace deterministic model. In many cases we will not get in trouble if we go down this route. For example, for most of the conditions one would likely put in a logistic model, one would lie in the smooth part of the curve (i.e., before $a = 1 + \sqrt{6}$). On the other hand, as we saw in looking at the logistic model for $a > 1 + \sqrt{6}$, if we consider a realistic model of a population, where the model must constantly be subject to noise, since that is the way the real world operates, then if we aggregate by simulating from the stochastic model as we did in Figure 8.6, we still obtain smooth, sensible results, even after the naive discrete logistic model has failed. But, as we have seen using two stochastic versions of examples from the work of Lorenz, there are times when the closed form is itself unreliable, even though the simulation-based aggregate, proceeding as it does from the microaxioms, is not. As rapid computing enables us to abandon more and more the closed form, we will undoubtedly find that simulation and stochastic modeling expand our ability to perceive the world as it truly is.

In 1993, Lorenz ([3], pp. 181–184) seriously addressed the question as to whether the flap of a butterfly's wings in Brazil could cause a tornado in Texas. He reckoned that isolation of the northern hemisphere from the southern in its wind currents might make it impossible, but within the same hemisphere, it was a possibility. In 1997, ([6], p. 360), remarking on his view that the weather is a chaotic phenomenon, Ian Stewart wrote, "Forecasts over a few days, maybe a week—that's fine. A month? Not a hope!"

If one listens closely to these statements, it is possible to hear the throbbing of the shamans' tom-toms celebrating the Festival of Unreason. Already, long-term models for the weather have been dramatically improved by noting the importance of driving currents in the jet stream. We now have the capability of forecasting whether winters will be warmer or colder than the norm in, say, New York. And high-frequency hurricane seasons are now being predicted rather well. It is large macroeffects—sunspots, El Niño, currents in the jet stream that drive the weather—not Brazilian butterflies. It is the aggregates which drive the weather. The tiny effects do not matter. In nature, smoothers abound, including noise itself.

Models for the forecasting of the economy have improved as well. Happily, rather than throwing up their hands at the futility of developing good models for real-world phenomena, a number of scientists are constantly drawing upon the scientific method for learning and making forecasts better. One of the bizarre traits of human beings is the tendency of some to believe models as stand-alones, unstressed by data. In Chapter 6 we noted how this can cause disaster, as in the case of the LTCM investment fund. On the other hand, some people refuse to look at models as having consequences for decision making. We note in Chapter 12 how this has characterized public health policy in the American AIDS epidemic. It is amazing how the same people can often take both positions simultaneously. For example, there are scientists who have worked hard to develop decision-free models for AIDS (this really takes some doing) and also work on trying to show how biological systems should be unstable (evidence to the contrary notwithstanding) due to the regions of instability in the discrete logistic model dealt with earlier in this chapter.

As time progresses, it is becoming ever more clear that naturally occurring realizations of mathematical chaos theory are difficult to find. The response of chaoticians is increasingly to include as "chaos" situations in which standard models for describing phenomena are being proved unsatisfactory or at least incomplete (see, e.g., [4]). Thus, "chaos" has been broadened to include "nonrobust" and "unstable." That some models are simply wrong and that others claim a completeness that is mistaken I do not question. The point being made in this chapter is that just because one can write down a chaotic model, it need not appear in nature, and that when it does we will probably view its effects only through the mediating and smoothing action of noise. Moreover, a philosophical orientation that we should give up on forecasting the weather or the economy because some-body has postulated, without validation, a chaotic model, is an unfortunate handshaking between the New Age and the Dark Ages.

A more complete analysis of chaos and noise is given in [7] and [8].

Problems

8.1. Let us consider the following birth process amongst single cell organ-isms. X is the population size, t the time, F the limiting population.

$$\text{Prob(birth in } [t, t + \Delta t)) \ = .001 X (F - X) \Delta t$$

Starting with an initial population of 10 at time 0, simulate a number of population tracks using a time increment of .01 and going to a time limit of 20. Take the time averages of 500 such tracks. Do you see evidence of chaotic behavior as you change F?

8.2. Again, consider the more general model

$$\text{Prob(birth in } [t, t + \Delta t)) \ = a X (F - X) \Delta t$$

Here a is positive, the initial population is 10. Again, you should satisfy yourself that, although there is considerable variation over time tracks, the average time track does not exhibit chaotic behavior.

8.3. Many of the older differential equation models of real world phenom-ena suffered from the necessity of not easily being able to incorporate the reality of random shocks. For example, let us consider the well known predator−prey model of Volterra ([5, 9]):

$$\frac{dx}{dt} \ = \ ax - bxy$$

$$\frac{dy}{dt} \ = \ -cy + cxy.$$

A simple Simpson's rule discretization works rather well:

$$x(t) \ = \ x(t - \Delta) + [ax(t - \Delta) - bx(t - \Delta)y(t - \Delta)]\Delta$$
$$y(t) \ = \ y(t - \Delta) + [-cy(t - \Delta) + dx(t)y(t - \Delta)]\Delta.$$

Here, the x are the fish and the y are the sharks. For $x(0) = 10$, $y(0) = 1$, $a = 3$, $b = 2$, $c = 2$, $d = 1.5$, $\Delta = .01$, we show the resulting plot for the first 13 units of time in Figure 8.11.

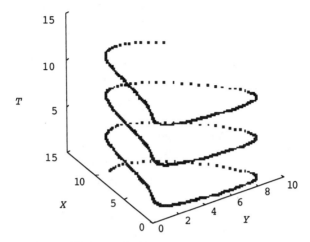

Figure 8.11. Volterra's Predator–Prey Model.

Such regularity is, of course, not supported by the evidence. The system is certainly stable, but not so regular. Show that if you change the discretization slightly to:

$$
\begin{aligned}
x(t) &= x(t - \Delta)(1. + 10z\Delta) + [ax(t - \Delta) - bx(t - \Delta)y(t - \Delta)]\Delta \\
y(t) &= y(t - \Delta) + [-cy(t - \Delta) + dx(t)y(t - \Delta)]\Delta,
\end{aligned}
$$

where z is a $\mathcal{N}(0, 1)$ random variable, then the stability of the system is maintained, but the path is much more realistically random.

8.4. One realization of the Henon attractor [5] is given, in discrete formulation, by

$$
\begin{aligned}
x_{n+1} &= y_n + 1 - 1.4x_n^2 \\
y_{n+1} &= 0.3x_n.
\end{aligned}
$$

For the Henon attractor, as with that of Lorenz, slight differences in the initial conditions produce great differences in the trajectories. Examine this fact and then observe the effect of adding a slight amount of noise to the system.

8.5. Chaos is frequently the concern of those who are worried about a data set somehow "on the edge" of stability. Consider, for example, a sample of size n from the Gaussian $\mathcal{N}(0, 1)$. Then, we know that if we compute the sample mean $\overline{X}$ and variance s^2, then

$$
t = \frac{\overline{X}}{s/\sqrt{n}}
$$

has a t distribution with $\nu = n - 1$ degrees of freedom. For $n = 2$, this is the Cauchy distribution which has neither expectation nor variance but is symmetrical about zero. For $n = \infty$, this is $\mathcal{N}(0, 1)$. Let us consider what happens when we sample from $t(\nu)$ adding on observations one at a time and computing the sequential sample mean:

$$\overline{T}_\nu(N) = \frac{(N - 1)\overline{T}_\nu(N - 1) + t_{\nu,N}}{N}$$

(a) Give plots of $\overline{T}_1(N)$ for N going from 1 to 5000.

(b) What happens if you throw away the 10% smallest observations and the 10% largest before computing $\overline{T}_1(N)$ (show this for $N = 10, 50, 100$ and 5000). This "trimming" is generally an easy way for minimizing the effects of long-tailed distributions.

(c) The Cauchy does not occur very often in nature (although it is easy to design a hardwired version: Just divide a $\mathcal{N}(0, 1)$ signal by another $\mathcal{N}(0, 1)$ signal; but that would not be a good design). Much more realistically, we carry out (a) but for $\nu = 3$.

References

[1] Lorenz, E.N. (1963). "Deterministic nonperiodic flow," *J. Atmospheric Sci.*, **20**, 130−141.

[2] Lorenz, E.N. (1964). "The problem of deducing the climate from the governing equations," *Tellus*, **16**, 1−11.

[3] Lorenz, E. N. (1993). *The Essence of Chaos*. London: UCL Press.

[4] Peters, E.E. (1994). *Fractal Market Analysis: Applying Chaos Theory to Investment and Economics*. New York: John Wiley & Sons.

[5] Pritchard, J. (1996). *The Chaos Cookbook: A Practical Programming Guide*. Oxford: Butterworth-Heinemann, 94, 166.

[6] Stewart, I. (1997). *Does God Play Dice?* London: Penguin Press.

[7] Thompson, J.R., Stivers, D.N., and Ensor, K.B. (1991). "SIMEST: a technique for model aggregation with considerations of chaos," in *Mathematical Population Dynamics*. O. Arino, D. Axelrod and M. Kimmel, eds. New York: Marcel Dekker, 483−510.

[8] Thompson, J.R. and Tapia, R.A. (1990). *Nonparametric Function Estimation, Modeling, and Simulation*. Philadelphia: SIAM, 244−252.

[9] Thompson, J.R. (1989). *Empirical Model Building*. New York: John Wiley & Sons, 71−76.

Chapter 9

Bayesian Approaches

9.1 Introduction

We carry out most of the work in this chapter based on one specific data set of Gehan and Freireich [4]. The convergence properties of the estimation procedures considered in this chapter, however, apply rather broadly. Those interested in convergence proofs are referred to Casella and George [1], Cowles and Carlin [2], or Tanner [6].

Let us consider a process in which failures occur according to an exponential distribution, that is,

$$F(t) = 1 - \exp(-\theta t). \tag{9.1}$$

Thus the probability no failure takes place on or before t is given by $\exp(-\theta t)$. Then, based on an independent sample of size n, the likelihood is given by

$$L(\theta) = \prod_{j=1}^{n} F'(t_j) = \theta^n \prod_{j=1}^{n} \exp(-\theta t_j) = \theta^n \exp\left(-\theta \sum_{j=1}^{n} t_j\right). \tag{9.2}$$

Then, the maximum likelihood estimator of θ can readily obtained be by taking the logarithm of the likelihood, differentiating with respect to θ, and setting the derivative equal to zero:

$$\log L(\theta) = n \log \theta - \theta \sum_{j=1}^{n} t_j \tag{9.3}$$

$$\frac{\partial \log L(\theta)}{\partial \theta} = \frac{n}{\hat{\theta}} - \sum_{j=1}^{n} t_j = 0. \tag{9.4}$$

This yields

$$\hat{\theta} = \frac{n}{\sum_{j=1}^{n} t_j}, \tag{9.5}$$

where the $\{t_j\}$ represent the n failure times.

Next, let us consider the case where one of the subjects did not yield an observed failure because the study ended at censoring time T. Then the likelihood becomes

$$
\begin{aligned}
L(\theta) &= \theta^{n-1} \prod_{j=1}^{n-1} \exp(-\theta t_j) \, \text{Prob (subject } n \text{ does not fail by } T) \\
&= \theta^{n-1} \prod_{j=1}^{n-1} \exp(-\theta t_j) \exp(-\theta T).
\end{aligned} \tag{9.6}
$$

Then the log likelihood becomes

$$\log L(\theta) = (n-1)\log(\theta) - \theta \sum_{j=1}^{n-1} t_j - \theta T, \tag{9.7}$$

yielding

$$\hat{\theta}_1 = \frac{n-1}{\sum_{j=1}^{n-1} t_j + T}. \tag{9.8}$$

9.2   The EM Algorithm

We note that our classical maximum likelihood estimator does not include any guess as to the when the nth subject might have failed had $T = \infty$. Yet, the times that $n-1$ individuals did fail does provide us with information relevant to guessing the nth failure time. Let us assume that our prior feelings as to the true value of θ are very vague: Essentially, we will take any value of θ from 0 to ∞ to be equally likely. Then, again, our log likelihood is given by

$$\log L(\theta) = n \log \theta - \theta \sum_{j=1}^{n-1} t_j - \theta t_n^*. \tag{9.9}$$

Now, the times of the $n-1$ failures are a matter of fact. The modal value of θ and the time of the unobserved failure time t_n^* are matters of inference. The EM algorithm is an iterative procedure whereby the hypothesized failure time t_n^* can be conjectured and the log likelihood reformulated for another attempt to obtain a new estimate for the modal value $\hat{\theta}$. Using our naive maximum likelihood estimator for θ, we have that the expected value for t_n^* is given by

$$
\begin{aligned}
t_{n,1}^* &= \frac{\int_T^\infty t_n^* \exp(-\hat{\theta}_1 t_n^*) dt_n^*}{\exp(-\hat{\theta}_1 T)} \\
&= T + \frac{1}{\hat{\theta}_1}.
\end{aligned}
\tag{9.10}
$$

Substituting this value in the log likelihood, we have

$$
\log L(\theta) = n \log \theta - \theta \sum_{j=1}^{n-1} t_j - \theta \left(T + \frac{1}{\hat{\theta}_1} \right).
\tag{9.11}
$$

This gives us

$$
\hat{\theta}_2 = \frac{n}{\sum_{j=1}^{n-1} t_j + \left(T + 1/\hat{\theta}_1 \right)}.
\tag{9.12}
$$

We then obtain a new expected value for the n'th failure time via

$$
t_{n,2}^* = T + \frac{1}{\hat{\theta}_2};
\tag{9.13}
$$

and so on.

In fact, for the exponential failure case, we can handle readily the more complex situation where n of the subjects fail at times $\{t_j\}_{j=1}^n$ and k are censored at times $\{T_i\}_{i=1}^m$. The m expectation estimates for $\{t_i\}_{i=1}^m$ are given at the kth step by $\{T_i + 1/\hat{\theta}_{k-1}\}$, and the log likelihood to be maximized is given by

$$
n \log \theta - \theta \left[\sum_{j=1}^{n-m} t_j + \sum_{i=1}^m (T_i + \frac{1}{\hat{\theta}_{k-1}}) \right].
\tag{9.14}
$$

Next, let us apply the EM in the analysis of the times of remission of leukemia patients using a new drug and those using an older modality of treatment. The data we use is from a clinical trial designed by Gehan and Freireich [4]. The database has been used by Cox and Oakes [3] as an example of the EM algorithm. Here we use it to examine the EM algorithm, data augmentation, chained data augmentation, and the Gibbs sampler.

Table 9.1. Leukemia Remission Times		
Ranked Survival	New Therapy	Old Therapy
1	6*	1
2	6	1
3	6	2
4	6	2
5	7	3
6	9*	4
7	10*	4
8	10	5
9	11*	5
10	13	8
11	16	8
12	17*	8
13	19*	8
14	20*	11
15	22	11
16	23	12
17	25*	12
18	32*	15
19	32*	17
20	34*	22
21	35*	23

Here (Table 9.1) an asterisk indicates that a patient's status was known to be remission until the (right censored time), at which time the status of the patient became unavailable. There is no pairing of patients in the two groups. We have simply ordered each of the two sets of 21 patients according to time of remission. Using (9.14) recursively to obtain θ for the new treatment, we have the results shown in Table 9.2.

The average survival time using the new therapy is $1/.025$, or 40 months. For the old therapy, average survival was only 8.67 months. So, the new therapy appears relatively promising. We note that in our use of the EM algorithm, we have not allowed our experience with the old therapy to influence our analysis of survival times of the new therapy.

We have chosen to explicate the EM algorithm by the use of a relatively simple example. It should be noted that, like the other algorithms we shall explore in this chapter, it performs effectively under rather general conditions. Clearly, the EM algorithm is, in fact, a data augmentation approach. So is the Gibbs Sampler. But the name *data augmentation algorithm* is generally reserved for the batch Bayesian augmentation approach covered in the next section.

Table 9.2. Iterations of EM Algorithm	
Iteration	θ
1	0.082569
2	0.041639
3	0.032448
4	0.028814
5	0.027080
6	0.026180
7	0.025693
8	0.025422
9	0.025270
10	0.025184
11	0.025135
12	0.025107
13	0.025091
14	0.025082
15	0.025077
16	0.025074
17	0.025072
18	0.025071
19	0.025070
20	0.025070

9.3 The Data Augmentation Algorithm

The EM procedure, although formally Bayesian, is, when one uses a diffuse prior, as we have done, an algorithm with which non-Bayesians generally feel comfortable. Many Bayesians, however, would be more comfortable with a procedure that gives the user, not simply the mode of a posterior distribution, but an estimate of the posterior distribution itself.

For example, let us suppose that the density function of a random variable X is given by $f(x; \theta)$, or, in Bayesian notation, $f(x|\theta)$. The joint density of a sample of x's of size n is then given by

$$\prod_{i=1}^{n} f(x_i|\theta).$$

Generally speaking, we will be interested in making inferences about the parameter θ in the light of a random sample $\{x_j\}_{j=1}^{n}$. Before we take any observations, we may well have some feelings about the parameter θ. Seldom will these feelings be so strong as to be of the sort, "We know that θ is precisely equal to 150.3741." If we were really so certain, why bother to collect data concerning the random variable X? It is much more likely that our feelings would be of the sort: "We are quite confident that θ is greater than 100 but less than 250." Expressing our prior feelings in terms of a prior distribution on the parameter space is not easy, for most people. Perhaps the major difficulty with a Bayesian approach is not on the basis of logic but on that of practicality. We may well have ideas about θ, absent any data. But it is not so easy to express these as a probability density function.

One way out of the difficulty is to require that the prior density function be such that the functional form will be unchanged by the addition of data. That is, the posterior distribution will have the same functional form as that of the prior density.

Let us return to the problem of exponentially distributed failure times. Here, we recall that

$$f(t_1, t_2, \ldots, t_n | \theta) = \theta^n \exp\left(-\theta \sum_{j=1}^{n} t_j\right).$$ (9.15)

Suppose that we decide to take as the prior density of θ, absent any data, a gamma density

$$p(\theta) = \frac{e^{-\lambda\theta} \lambda^\alpha \theta^{\alpha-1}}{\Gamma(\alpha)}.$$ (9.16)

Then

$$E(\theta) = \alpha/\lambda$$ (9.17)
$$\text{Var}(\theta) = \alpha/\lambda^2.$$ (9.18)

It is not unreasonable to suppose that we have some notion as to our prior feelings as to the mean and variance of θ. These feelings enable ready guesses as to appropriate values of λ and α. Furthermore, we can then write down the joint density of θ and the failure times as

$$p(\theta; t_1, t_2, \ldots, t_n) = p(\theta) f(t_1, t_2, \ldots, t_n | \theta)$$
$$v = \frac{\theta^{n+\alpha-1} \lambda^\alpha \exp[-\theta(\lambda + \sum t_j)]}{\Gamma(\alpha)}.$$ (9.19)

If we then obtain the marginal density of $t_1, t_2, \ldots, t_n$, $h(t_1, t_2, \ldots, t_n)$, for example, we can obtain the posterior density of θ, via

$$p(\theta | t_1, t_2, \ldots, t_n) = \frac{p(\theta; t_1, t_2, \ldots, t_n)}{h(t_1, t_2, \ldots, t_n)}.$$ (9.20)

Here we have

$$h(t_1, t_2, \ldots, t_n) = \frac{\lambda^\alpha}{\Gamma(\alpha)} \int_0^\infty \theta^{i+m-1} \exp\left[-\theta(\lambda + \sum t_j)\right] d\theta$$
$$= \frac{\Gamma(m+n)}{\Gamma(\alpha)} \frac{\lambda^\alpha}{(\lambda + \sum t_j)^{n+\alpha}}.$$ (9.21)

Then, readily, we have

$$p(\theta|t_1, t_2, \ldots, t_n) = \frac{(\lambda + \sum t_j)^{n+\alpha}}{\Gamma(\alpha + n)} \theta^{n+\alpha-1} \exp\left[-\theta(\lambda + \sum t_j)\right]. \quad (9.22)$$

But this density is also of the gamma form with

$$\lambda^* = \lambda + \sum t_j$$

and

$$\alpha^* = \alpha + n,$$

that is,

$$p(\theta|t_1, t_2, \ldots, t_n) = \frac{e^{-\theta\lambda^*}(\lambda^*)^{\alpha^*}\theta^{\alpha^*-1}}{\Gamma(\alpha^*)}, \quad (9.23)$$

where

$$E(\theta|t_1, t_2, \ldots, t_n) = \frac{\alpha^*}{\lambda^*} = \frac{n+\alpha}{\lambda + \sum t_j} \quad (9.24)$$

and

$$\text{Var}(\theta|t_1, t_2, \ldots, t_n) = \frac{\alpha^*}{(\lambda^*)^2} = \frac{n+\alpha}{(\lambda + \sum t_j)^2}. \quad (9.25)$$

In the case where some of the failure times were missing due to censoring, we could use the EM algorithm to find an improved estimate of the mode of $p(\theta|t_1, t_2, \ldots, t_n)$. We shall, however, consider a strategy for estimating the posterior density itself, either through obtaining a knowledge of the function $p(\theta|t_1, t_2, \ldots, t_n)$ or a pointwise (in θ) evaluator of $p(\theta|t_1, t_2, \ldots, t_n)$.

Let us suppose we observe the failure times $t_1, t_2, \ldots, t_{n-m}$ but are missing $\{t_j\}_{j=n-m+1}^n$, since these individuals were lost from the study at times $\{T_j\}_{j=n-m+1}^n$. The *data augmentation algorithm* proceeds as described below:

Data Augmentation Algorithm

1. Sample θ_j from $p(\theta_j|t_1, t_2, \ldots, t_n)$.

2. Generate $t_{n-m+i}, \ldots, t_n$ from $\theta_j \exp(-\theta_j t_{n-m+i})$ (with the restriction that $t_{n-m+i} \geq T_{n-m+i}$).

3. Repeat Step 2 N times.

4. Compute $\bar{T} = 1/N \sum_{i=1}^N \sum_{j=1}^n t_{ji}$.

5. Let $\lambda^* = \lambda + \bar{T}$.

6. Let $\alpha^* = \alpha + n^*$.

7. Then the new iterate for the posterior distribution for θ is given by

$$p(\theta|t_1, t_2, \ldots, t_n) = \frac{e^{-\theta\lambda^*}(\lambda^*)^{\alpha^*}\theta^{\alpha^*-1}}{\Gamma(\alpha^*)}.$$

8. Return to Step 1, repeating the cycle M times or until the estimates λ^* and α^* stabilize.

For the first pass through the cycle, we use for $p(\theta|t_1, t_2, \ldots, t_n)$, simply the prior density for θ, $p(\theta)$. Typically, N is quite large, say on the order of 1000. Again, typically, we will go through the repeat cycle until the estimates of the posterior distribution for θ stabilize.

We note that, unlike the EM algorithm, there is no maximization step in data augmentation, rather a series of expectations. It is interesting to note under rather general conditions, data augmentation does stabilize to a "fixed point" (in function space) under expectation iterations.

Let us consider using the data augmentation algorithm on the remission data for the new treatment in Table 9.1. First of all, we need to ask whether there is a reasonable way to obtain the parameters for the gamma prior distribution of θ. In clinical trials, there is generally the assumption that the newer treatment must be assumed to be no better than the old treatment. So, in this case, we might consider obtaining estimates of the two parameters by looking at the data from the older (control) therapy.

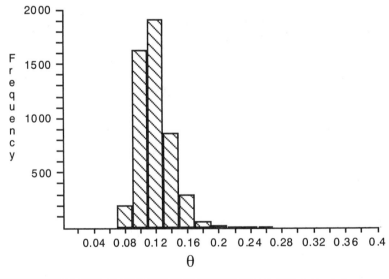

9.1. Resampled Values of θ.

Let us use a simple bootstrap approach to achieve this objective. Sampling with replacement 5000 times from the 21 control remission times (we have used the *Resampling Stats* package [5], but the algorithm can easily be programmed in a few lines of Fortran or C code). Now, for each of the runs, we have computed an estimate for θ, which is simply the reciprocal of the average of the remission times. A histogram of these remission times is given in Figure 9.1.

For the control group, the bootstrap average of θ is .11877 and the variance is .00039018. Then, from (9.17) and (9.18), we have as estimates for λ and α, 304.40 and 36.15, respectively. Returning to the data from the new drug, let us use $N = 1000$.[1] After 400 simulations, we find that

$$\bar{T} = 1/400 \sum_{i=1}^{400} \sum_{j=1}^{12} t_{ji} = 539.75 \tag{9.26}$$

$$\begin{aligned} \alpha^* &= 36.15 + 21 = 57.15 \\ \lambda^* &= 304.4 + 539.75 = 844.15. \end{aligned} \tag{9.27}$$

Computing the posterior mean and variance of θ, we have

$$E(\theta|\bar{T}) = \frac{\alpha^*}{\lambda^*} = .0677;$$

$$\text{Var}(\theta|\bar{T}) = \frac{\alpha^*}{\lambda^{*2}} = .0000802. \tag{9.28}$$

Essentially, we can approximate the posterior distribution of θ as being Gaussian with mean .0677 and variance .0000802.

It seems that waiting for such a large number (1000) of simulations to update our estimates for λ^* and α^* may be somewhat inefficient. Above we have used $N = 1000$ and $M = 400$.

Now, we need not use a large value for N in the generation of plausible t_{n-m+j}. In fact, by using a large value for N, we may be expending a large amount of computing time generating $t_{n-m+1}, t_{n-m+2}, \ldots t_n$ values for θ values generated from posterior distribution estimates which are far from the mark. Let us go through the data augmentation algorithm with $N = 50$ and $M = 10,000$.[2] This requires approximately the same number of computations as the first run ($N = 1000$ and $M = 400$) run. We now have the estimates for the posterior mean and variance of θ being .06783 and .0000848, respectively. Clearly, the use of a smaller value for N has not changed our estimate for the parameters for the posterior distribution of θ. In Figure 9.2, we show a histogram of the θ values generated for the 10,000

---

[1] The author wishes to thank Patrick King and Mary Calizzi for the computations in the first data augmentation run.

[2] The author wishes to thank Otto Schwalb for the computations and graphs in the balance of this chapter.

runs. On this, we superimpose two Gaussian distributions with variance .0000802, one centered at the mean of the generated θ's and the other at the median.

Data Augmentation

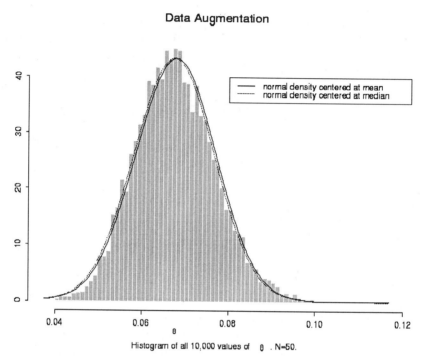

Histogram of all 10,000 values of θ. N=50.

Figure 9.2. Histogram of θ.

It could be argued that we might, in fact, advocate recomputing our guess as to the true posterior distribution after each generation of a set of missing values. This is the *chained data augmentation algorithm*. We note that, now, when we go from Step 1 to Step 8, we "transit" from one value of θ to another. We also observe the one step memory of the process. Knowing $p(\theta_j | t_1, t_2, \ldots, t_n)$ is sufficient to generate $t_{n-m+1}, \ldots, t_n$, which, in turn, is sufficient to generate $p(\theta_{j+1} | t_1, t_2, \ldots, t_{nj}^*)$. Thus, the chained data augmentation algorithm is Markovian. And, clearly, the more general data augmentation algorithm is Markovian as well. In the present example, using $N = 1$ (i.e., the chained data augmentation algorithm) took less than 5% of the running time of the data augmentation algorithm with $N = 1000$ and $M = 400$ (with essentially the same results).

At the level of intuition, it would appear that the only reason to use an N greater than 1 would be to guarantee some sort of stability in the estimates. It turns out that this is not necessary, and this fact leads us immediately to

the dominant simulation-based paradigm by which orthodox Bayesians deal
with missing values, namely, the *Gibbs Sampler*, which subject we address
in the following section.

9.4   The Gibbs Sampler

Next, let us consider the situation where we decide to model failure times
according to the normal distribution

$$f(t_1, t_2, \ldots, t_n | \mu, h^2) = \left(\frac{h^2}{2\pi}\right)^{n/2} \exp\left[-\frac{1}{2}h^2 \sum_{j=1}^{n}(t_j - \mu)^2\right], \qquad (9.29)$$

where both μ and $h^2 = 1/\sigma^2$ are unknown. We start with a data set consisting of failure times $t_1, t_2, \ldots, t_{n-m}$ but are missing $t_{n-m+1}, t_{n-m+2}, \cdots, t_n$.
For the most recent estimates for μ and h^2 we shall generate surrogates for
the m missing values from a Gaussian distribution, with these estimates
for μ and h^2 imposing the restriction that we shall, in the generation of
t_{n-m+j}, say, discard a value less than the censoring time T_{n-m+j}.

The natural conjugate prior here is

$$p(\mu, h^2 | M', V', n', \nu')$$

$$= K \exp\left[-\frac{1}{2}h^2 n'(\mu - M')^2\right] \sqrt{h^2} \exp[-\frac{1}{2}h^2 V'\nu'](h^2)^{\nu'/2-1}, \qquad (9.30)$$

where K is a constant of integration. (We shall, in the following, use K
generically, i.e., one symbol, K will be used for all constants of integration.)
Let

$$M = \frac{1}{n}\sum_{i=1}^{n}t_i$$

$$\nu = n - 1$$

$$V = \frac{1}{\nu}\sum_{i=1}^{n}(t_i - M)^2.$$

Then the posterior distribution is given by

$$p(\mu, h^2 | M'', V'', n'', \nu'')$$

$$= K \exp\left[-\frac{1}{2}h^2 n''(\mu - M'')^2\right] \sqrt{h^2} \exp\left[-\frac{1}{2}(h^2)V''\nu''\right](h^2)^{\nu''/2-1}, \quad (9.31)$$

where

$$n'' = n' + n$$

$$M'' = \frac{1}{n''}(n'M' + nM)$$

$$\nu'' = \nu' + (n - 1) + 1$$

$$V'' = \frac{1}{\nu''}[(\nu'V' + n'M'^2) + (\nu V + nM^2 - n''M''^2)].$$

Immediately, then, we have the possibility of using the *chained data augmentation algorithm* via

1. Generate (μ, h^2) from $p(\mu, h^2 \mid M'', V'', n'', \nu'')$.

2. Generate $\{t_{n-m+j}\}_{j=1}^m$ from

$$f(t_{n-m+j} \mid \mu, h^2) = \left(\frac{h^2}{2\pi}\right)^{1/2} \exp\left[-\frac{h^2}{2}(t_{n-m+j} - \mu)^2\right],$$

   where $t_{n-m+j} > T$.

3. Return to Step 1.

Such a strategy is rather difficult to implement, since it requires the generation of two-dimensional random variables (μ, h^2). And we can well imagine how bad things can get if the number of parameters is, say, four or more. We have seen in Chapters 1 and 2 that multivariate random number generation can be unwieldy. There is an easy way out of the trap, as it turns out, for according to the *Gibbs sampler* paradigm, we simply generate from the one-dimensional distributions of the parameters sequentially, conditional upon the last generation of the other parameters and the missing value data. Let us see how this is done in the case of the current situation.

The posterior density for h^2 (conditional on the data including the surrogates for the missing failure times) is given by

$$p_{h^2}(h^2 \mid V'', \nu'') = K \exp\left(-\frac{yV''\nu''}{2}\right) y^{\nu''/2 - 1}. \tag{9.32}$$

We may then obtain the conditional density for μ given h^2 and the data by

$$p_{\mu|h^2}(\mu \mid h^2, M'', V'', n'', \nu'') = \frac{p(\mu, h^2 \mid M'', V'', n'', \nu'')}{p(h^2 \mid V'', \nu'')}$$

$$= K \exp\left[-\frac{1}{2}h^2 n''(\mu - M'')^2\right]. \tag{9.33}$$

Similarly, the posterior density for μ (conditional on the data including the surrogates for the missing failure times) is given by

$$p_\mu(\mu \mid M'', n'', V'', \nu'') = K\left[\nu'' + \frac{(\mu - M'')^2 n''}{V''}\right]^{-(\nu''+1)/2}. \tag{9.34}$$

We may then find the conditional distribution for h^2 given μ and the data by

$$p_{h^2|\mu}(h^2|\mu, M'', V'', n'', \nu'')$$

$$= \frac{p(\mu, h^2|\ M'', V'', n'', \nu'')}{p_\mu(\mu|M'', n'', V'', \nu'')} \qquad (9.35)$$

$$= K(h^2)^{\frac{\nu+1}{2}} e^{[-\frac{1}{2}(n''(\mu-M'')^2+V''\nu'')h^2]}.$$

In summary, the missing failure times are generated from a normal distribution with the current estimates of mean $= \mu$ and variance $= 1/h^2$. μ, conditional on the generation both of data including missing values for failures and h^2, is generated from a normal distribution with mean M'' and variance $1/(h^2n'')$. h^2, conditional on the data and pseudodata and μ, is a χ^2 variable with $\nu''+1$ degrees of freedom divided by $n''(\mu-M'')^2+V''\nu''$. Clearly, such one-dimension-at-a-time samplings are extremely easy.

Let us suppose we observe the failure times $t_1, t_2, \ldots, t_{n-m}$ but are missing $\{t_j\}_{j=n-m+1}^{n}$, since these individuals were lost from the study at times $\{T_j\}_{j=n-m+1}^{n}$. The *Gibbs sampler algorithm* proceeds thusly: At the start, we shall set $M'' = M'$, $V'' = V'$, $n'' = n'$, $V'' = V'$.

1. Generate μ from $p_\mu(\mu|M'', n'', V'', \nu'')$.

2. Generate h^2 from $p_{h^2}(h^2|V''\nu'')$.

3. Generate $\{t_{n-m+j}\}_{j=1}^{m}$ from

$$f(t_{n-m+j}|\mu, h^2) = \left(\frac{h^2}{2\pi}\right)^{1/2} \exp\left[-\frac{h^2}{2}(t_{n-m+j} - \mu)^2\right].$$

where $t_{n-m+j} > T$.

4. Return to Step 1.

By simply recording all the μ and h^2 (hence σ^2), recorded in sequence of generation, over, say, 10,000 iterations, we can obtain a histogram picture of the joint posterior density for (μ, h^2).

We can do more. Suppose that

$$\bar{M}'' = \text{Ave}\{M''\}$$
$$\bar{V}'' = \text{Ave}\{V''\}.$$

Then we have

$$p(\mu, h^2|t_1, t_2, \ldots, t_n) \approx p(\mu, h^2|\bar{M}'', \bar{V}'', n'', \nu''). \qquad (9.36)$$

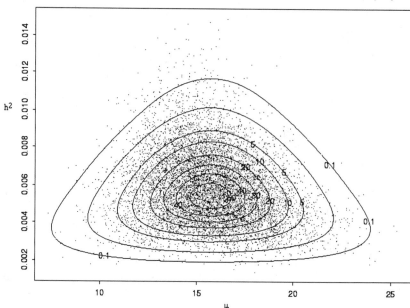

Figure 9.3. Posterior Draws of (μ, h^2)

For the Gehan−Freireich data, we have $n = 21$. Using the control group to obtain the initial parameters of the prior, we have $M' = 8.667$, via a bootstrap sample of size 5000. Looking at the reciprocals of the variances of 5000 bootstrap samples of size $n' = 21$ from the control group, we have $E(h) = .028085$ and $\text{Var}(h) = .00012236$. This gives

$$V' = \frac{1}{E(h)} = 35.606$$

$$\nu' = 2\frac{E(h)^2}{\text{Var}(h)} = 2\frac{(.028085)^2}{.00012236} = 12.89.$$

Performing 10,000 samples of μ and h generated one after the other (with the first 1000 discarded to minimize startup effects), we arrive at

$$
\begin{aligned}
M'' &= 15.716 \\
V'' &= 179.203 \\
n'' &= n' + n = 42 \\
\nu'' &= 12.89 + 21n = 33.89.
\end{aligned}
$$

The results of these samplings of (μ, h^2) are shown in Figure 9.3. The marginal histograms of μ and h^2 are given in Figures 9.4 and 9.5 respectively.

Histogram of 9,000 Generations of μ with p(μ| M''=15.716, V'' = 179.203, n'' = 42, υ'' = 33.89) Superimposed

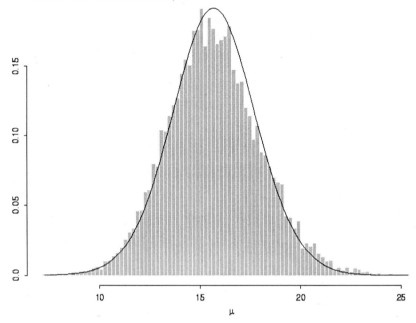

Figure 9.4. Posterior Draws of μ.

Histogram of 9,000 Generations of h² with p $_{h^2|\mu}$ (h²| M''=15.716, V'' = 179.203, n'' = 42, υ'' = 33.89) Superimposed

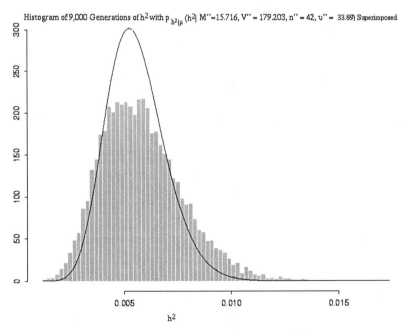

Figure 9.5. Posterior Draws of h^2.

By way of comparison with the earlier data augmentation approach taken in Figure 9.4, in Figure 9.6, we show a histogram taken from the values of 450 but coded as $\theta = 1/\mu$. We note that Figure 9.4 and Figure 9.6 are essentially the same. The average survival posterior mean for the survival time, in both cases is roughly 15.7 months.

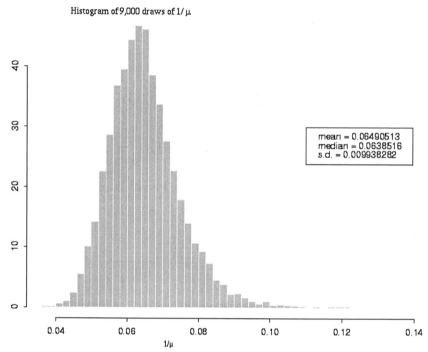

Figure 9.6. **Posterior Draws of** $1/\mu$

9.5   Conclusions

We have noted that the data augmentation and the Gibbs Sampler procedures employed both give, for the new treatment, approximately the same posterior mean (15.7 months) for the average survival time. This value is roughly twice as long as the average survival time for the old treatment (8.7 months). But in the case of the EM algorithm, our estimate for the average survival time was a much more optimistic 40 months. What is the reason for the discrepancy?

The discrepancy occurs because in the data augmentation and Gibbs Sampler procedures, we elected to use a prior distribution for the parameters of interest which was obtained from data using the old treatment. In the EM algorithm, on the other hand, we used a "noninformative" prior distribution (i.e., one with a very large variance).

Philosophically, some Bayesians would object to our utilization of the results from the old treatment to estimate the parameters of the prior distribution for the new treatment. It could be argued that such a step was unduly pessimistic and smacked of an underlying frequentist mindset whereby the prior distribution was formed with a de facto null hypothesis in mind, namely that the new therapy produced no better average survival times than the old. A Bayesian statistician would probably prefer to consult a panel of advocates of the new treatment and using their insights, attempt to obtain the parameters of the appropriate prior distribution.

In the long run, after the new treatment has been employed on many individuals, it will not make much difference what (nondegenerate) prior distribution we elected to use. But a standard Bayesian claim is that Bayesian techniques enable a more rapid change from a less effective to a more effective procedure. In the case considered here, we have used what Bayesian would consider a rather pessimistic prior. This is unusual in Bayesian analysis. Most Bayesians would tend to use a prior no more pessimistic than the noninformative prior which we used in the case of the EM algorithm. And such a prior is generally roughly equivalent to a standard nonBayesian, frequentist approach. Thus, there is kind of practical bias in favor of the new therapy in most Bayesian analyses. If there is physician opinion indicating the new treatment is much better, then that is incorporated into the prior. But, absent good news, the statistician is supposed to default to a noninformative prior. That may indeed give a running start to those wishing to change the protocol.

In the problems, we give examples of a variety of possible prior assumptions which might be used.

Problems

9.1. Consider the situation where a random variable X has the normal distribution $\mathcal{N}(\theta, 1)$ and θ has the normal distribution $\mathcal{N}(0, 1)$. Create a sample of size 1000 by first sampling a θ and then an X 1000 times.
(a) Create a two-dimensional histogram of (θ, X).
(b) Create a one-dimensional histogram of the marginal density of X.
(c) Find explicitly the marginal density of X.
9.2. We recall that a Poisson random variable has the probability function

$$P(X) = e^{-\theta}\frac{\theta^X}{X!}.$$

Suppose that θ has the exponential density

$$f(\theta) \quad = \quad e^{-\theta} \text{ for } \theta \geq 0$$
$$= \quad 0 \text{ for } \theta < 0.$$

Generate first a θ, then an X. Do this 1000 times.

(a) Show a two-dimensional histogram of (θ, X).

(b) Create a one-dimensional histogram of the probability function of X.

(c) Find the one-dimensional probability function of X.

The time that it takes for a member of a group of small inner city businesses in a particular city to develop credit problems seems to be short. The lending agency is considering making an attempt to increase the time to failure by giving companies free counseling at monthly intervals. Two subgroups of size 25 each have their times until first credit difficulty recorded. The members of the first group (A) do not receive the counseling. The members of the second group (B) do. Below, we show the results of the first two years. The times (in months) till time of first credit problem are given in Table 9.3.

Table 9.3. Time Until First Credit Problem		
Rank	Group A	Group B
1	0	1
2	1	1
3	1	1
4	1	4
5	1	4
6	1	4
7	2	5
8	2	5
9	2	6
10	3	7
11	5	8
12	5	8
13	7	10
14	7	11
15	8	12
16	9	12
17	10	14
18	11	16
19	12	24*
20	14	24*
21	15	24*
22	20	24*
23	24*	24*
24	24*	24*
25	24*	24*

9.3. Assuming that the time to first problem is exponentially distributed, use the EM algorithm to estimate θ_A and θ_B.

9.4. One useful Bayesian approach in deciding whether to make the counseling a standard protocol would be to compute the posterior distributions of θ_A and θ_B. Do this, utilizing the data augmentation algorithm developed in Section 9.3.

9.5. Next, let us consider the situation where we decide to model failure times according to the normal distribution

$$f(t_1, t_2, \ldots, t_n | \mu, h^2) = \left(\frac{h^2}{2\pi}\right)^{n/2} \exp\left[-\frac{1}{2}h^2 \sum_{j=1}^{n}(t_j - \mu)^2\right],$$

where both μ and $h^2 = 1/\sigma^2$ are unknown. Using the approach in Section 9.4, obtain Gibbs sampler estimates for the posterior distributions of (μ_A, h_A^2) and (μ_B, h_B^2).

9.6. As has been mentioned earlier in this chapter, the results for finding an estimate for θ in the case of the Gehan–Freireich leukemia data are similar for the data augmentation and the Gibb's sampler procedures. But these results are quite different from those obtained by the use of the EM algorithm. This would appear to be due to the fact that in the case of our data augmentation and Gibb's sampler analyses we have used a prior distribution based on the survival results of the old therapy. Moreover, the sample size of the old therapy was the same as that of the new therapy. We noted that the bootstrap estimate for θ based on the old therapy was .11877. The EM estimate for θ was .02507. The average of these two estimates is .07192, a figure which is roughly similar to the estimate we obtained with data augmentation, .0677. It could be argued, therefore, that by using the old procedure to obtain a prior density for θ, we have swamped the good results of the new therapy with the poorer results of the old therapy.

We recall from equation (9.22) that the posterior distribution of θ is given by

$$p(\theta | t_1, t_2, \ldots, t_n) = \frac{e^{-\theta\lambda^*}(\lambda^*)^{\alpha^*}\theta^{\alpha^*-1}}{\Gamma(\alpha^*)},$$

where

$$E(\theta | t_1, t_2, \ldots, t_n) = \frac{\alpha^*}{\lambda^*} = \frac{n+\alpha}{\lambda+\sum t_j}$$

and

$$\text{Var}(\theta | t_1, t_2, \ldots, t_n) = \frac{\alpha^*}{(\lambda^*)^2} = \frac{n+\alpha}{(\lambda+\sum t_j)^2}.$$

Suppose that we decide to use our resampled value of the prior distribution's mean using the data from the old therapy (.11877), but decide that we wish to increase dramatically the estimate for the variance of the prior beyond the old therapy resampled value (.00390). Go through the data augmentation algorithm using higher and higher values for the variance of the prior distribution for θ and see what the data augmentation posterior means for θ are. One might expect that as the variance of the prior increases without limit, the data augmentation and Gibb's sampler results would look much more like those obtained by the EM algorithm. Is this the case?

References

[1] Casella, G. and George, E.I. (1992). "Explaining the Gibbs Sampler," *Amer. Statist.*, 167–174.

[2] Cowles, M.K. and Carlin, B.P. (1996). "Markov chain Monte Carlo convergence diagnostics: a comparative review," *J. Amer. Statist. Assoc.*, 883–904.

[3] Cox, D.R., and Oakes, D. (1984). *Analysis of Survival Data*, **8**, London: Chapman & Hall, 166–168.

[4] Gehan, E.A. (1965). "A generalized Wilcoxon test for comparing arbitrarily single-censored samples," *Biometrika*, 203–23.

[5] Simon, J.L.(1990). *Resampling Stats*. Arlington, Va.: Resampling Stats, Inc.

[6] Tanner, M.A. (1993). *Tools for Statistical Inference*. New York: Springer-Verlag.

Chapter 10

Resampling-Based Tests

10.1 Introduction

One of the earliest data sets considered in extenso during the modern statistical era is that of cross-fertilized versus self-fertilized *Zea mays* (a.k.a. *corn*) submitted by Charles Darwin to his kinsman, Francis Galton. The analysis of these data by Ronald Fisher [6] in 1935 reveals many of the problems of hypothesis testing.

Philosophically speaking, perhaps the most significant issue is not that of the underlying probability distribution(s) of the data. Rather, it is crucial that we are able to state with some specificity that which we are attempting to test, including both the (generally status quo) *null hypothesis* and the *alternative hypothesis*. It is intriguing that Fisher tended to dismiss fears that the investigator would be ruined by making assumptions of distributional normality. He felt that to go nonparametric would be the more devastating step. Subsequent experience indicates that it is frequently the case that no great loss of efficiency is to be experienced if we attempt to utilize some simulation-based resampling strategy as an alternative to one based on the assumption of Gaussianity of the data. No doubt, for small samples there is frequently a real price to be paid. However, we live in increasingly data-rich times. The gains we experience in robustness for the resampling strategies are generally well worth the modest price we may pay for slight loss of power vis-à-vis the classical normal theory–based tests favored by Fisher. Much of Fisher's work actually points toward resampling tests, and Fisher lived into the computer age. Computing potential, however, seems to lead useful application by decades. The world would have to wait two decades after the death of Fisher until Bradley Efron's discovery of the bootstrap.

In the case of Darwin's corn data, we have a number of possible questions to be addressed. Naturally, although Darwin was looking specifically at a certain variety of a specific grain plant, the implicit question is more general: Is cross-fertilization good or bad? The common wisdom of most cultures is that it is probably a good thing. And utilizing almost any kind of analysis on the height based Darwin corn data, we arrive at an answer consistent with the common wisdom.

In the case of Darwin and Galton and Fisher, we see that the surrogate for "goodness" is stalk height at a fixed time after sprouting. It could have been otherwise. Darwin might have used other criteria (e.g., grain yield, resistance to disease and drought, flavor of grain, etc.). We cannot include every possible criterion. The sociologist/anthropologist Ashley Montague noted that by a number of criteria, sponges were superior to human beings. The selection of a criterion to be measured and analyzed almost always requires a certain amount of subjectivity. We always hope that the criterion selected will be positively correlated with other, perhaps relevant but unnoticed criteria.

In the case of the Darwin corn data, the null hypothesis selected is that there is no difference in stalk height between the cross-fertilized and self-fertilized plants. But when it comes to the alternative hypothesis the specificity is more vague. For example, we could select an alternative hypothesis model (à la Darwin, Galton, and Fisher) that cross-fertilization always increases stalk height with variation from this rule being due to unexplained factors. Or, we could hypothesize that, on the average, cross-fertilization leads to increased stalk height. Or, we might opine that the median stalk height of a group of cross-fertilized plants tends to be greater than that of a group of self-fertilized plants. Each of these alternative hypotheses is different (although the first implies the next two). In the case of the Darwin data, each of the hypotheses appears to be supported by the data.

Selection of hypotheses is not easy, and the literature is replete with examples of studies where inappropriate hypotheses led to ridiculous or pointless conclusions. But an ivory tower disdain of specifying hypotheses leads, as a practical matter, to the radical position of Ashley Montague, where Shakespeare is no better than a sponge. That is more multiculturalism than is consistent with progress, scientific or otherwise.

10.2   Fisher's Analysis of Darwin's Data

In Table 10.1 we show the data essentially as presented by Darwin to Galton. We note here a "pot effect" and, possibly, an "adjacency within pot effect." At any rate, Darwin presented the data to Galton paired. Naturally, Darwin had made every attempt to equalize soil and water conditions across pots. It might well appear to us, in retrospect, that such equalization should be readily obtainable and that we might simply pool the data into 15 cross-fertilized and 15 self-fertilized plants.

Table 10.1			
Pot	Crossed	Self-Fertilized	Difference
I	23.500	17.375	6.125
I	12.000	20.375	-8.375
I	21.000	20.000	1.000
II	22.000	20.000	2.000
II	19.124	18.375	0.749
II	21.500	18.625	2.875
III	22.125	18.625	3.500
III	20.375	15.250	5.125
III	18.250	16.500	1.750
III	21.625	18.000	3.625
III	23.250	16.250	7.000
IV	21.000	18.000	3.000
IV	22.125	12.750	9.375
IV	23.000	15.500	7.500
IV	12.000	18.000	-6.000

Let us first consider Fisher's main analysis based on the differences shown in the fourth column. Fisher made the assumption that the stalk heights were normally distributed. Let us assume that the height of a stalk is due to two factors:

1. Cross-fertilized (Y) versus self-fertilized (X)

2. pot effect (I, II, III, or IV).

Then if Y is $\mathcal{N}(\mu_Y, \sigma_Y^2)$ and X is $\mathcal{N}(\mu_X, \sigma_X^2)$, let us consider the differences:

$$
\begin{aligned}
d_A &= Y + \text{Pot}_A - (X + \text{Pot}_A) \\
&= Y - X. \tag{10.1}
\end{aligned}
$$

and similarly for the other differences. So, d is $\mathcal{N}(\mu_Y - \mu_X, \sigma_Y^2 + \sigma_X^2)$. Then, if we look at the 15 differences, we can obtain

$$
\bar{d} = \frac{1}{15} \sum_{j=1}^{15} d_j \tag{10.2}
$$

and

$$
s^2 = \frac{1}{14} \sum_{j=1}^{15} (d_j - \bar{d})^2. \tag{10.3}
$$

and perform a t-test with 14 degrees of freedom. The t value is 2.148. Accordingly, the null hypothesis that

$$
H_0 : \mu_Y = \mu_X
$$

is rejected at the .05 level.

Now Fisher did a bit of combinatorial work on the data that was presented to Galton and came up with a clever nonparametric test. He noted that if the null hypothesis were true, there was an equal chance that each one of the pairs came from the self-fertilized or the cross-fertilized. He noted that in the way the pairs were presented, the sum of the differences was 39.25 inches. Looking over all the possible swaps in each pair, he noted that there were $2^{15} = 32,768$ possibilities. In only 863 of these cases was the sum of the differences as great as 39.25. Thus, the null hypothesis would be rejected at the $2 \times 863/32,768 = .0526$ level.

Fisher was extremely pleased that his nonparametric test agreed with the result of his t-test. But he was hardly a champion of nonparametric tests as alternatives to normal theory-based ones:

> The utility of such nonparametric tests consists in their being able to supply confirmation whenever, rightly or, more often, wrongly, it is suspected that the simpler tests have been appreciably injured by departures from normality.
>
> They assume less knowledge, or more ignorance, of the experimental material than does the standard test, and this has been an attraction to some mathematicians who often discuss experimentation without personal knowledge of the material. In inductive logic, however, an erroneous assumption of ignorance is not innocuous: it often leads to manifest absurdities. Experimenters should remember that they and their colleagues usually know more about the kind of material they are dealing with than do the authors of text-books written without such personal experience, and that a more complex, or less intelligible, test is not likely to serve their purpose better, in any sense than those of proved value in their own subject ([6], p. 48).

The implication here is that by removing the assumption of normality, we will come up with a test that is less powerful than the t-test; that is, we will be less likely to reject the null hypothesis when it is, in fact, false. A fair assumption for his time, but experience has led us to have some disagreement with Fisher's seemingly plausible implication.

First, the t-test has a natural robustness against departures from normality of the data. Why? Because it deals with $\bar{x}$. If the sample size is large enough (and sometimes a sample size of 5 will do nicely), then the central limit theorem will push $\bar{x}$ towards near normality. Other tests, such as the $\mathcal{F}$-test are much more fragile.

Second, experience indicates that rank tests, for example, have good efficiency, relative to their parametric substitutes, even when the data is normal. And when the data is not drawn from a normal distribution, these tests can give useful results when the normal theory-based tests fail badly.

Nevertheless, we must agree that, particularly with multivariate data, normality is an outstanding bonus, which can buy us a great deal. So, we can agree that it is not unreasonable to follow Fisher's suggestion, when possible, namely to use nonparametric tests and normal theory-based tests in parallel, always happy when the assumption of normality turns out to be reasonable.

Let us consider, in the case of the Darwin data, a popular nonparametric test, namely the Wilcoxon signed rank test. Let us suppose that we have n differences of variables, which we suppose, under the null hypothesis, to be taken from the same distribution. Here is the Wilcoxon protocol:

Wilcoxon Signed Rank Test

1. Rank the absolute differences from 1 to n.

2. Multiply each of these ranks by the sign of the difference.

3. Sum these signed ranks to give the statistic W.

4. Consider the signed rank of the first difference R_1.

5. Clearly,
$$E(R_1) = -\frac{n+1}{2} \times \frac{1}{2} + \frac{n+1}{2} \times \frac{1}{2} = 0 \; ;$$

$$\text{Var}(R_1) = E(R_1^2) = \frac{1^2 + 2^2 + \ldots + n^2}{n} = \frac{(n+1)(2n+1)}{6}.$$

6. Then, under the null hypothesis,
$$E(W) = \sum_{i=1}^{n} R_i = 0;$$

$$\text{Var}(W) = \frac{n(n+1)(2n+1)}{6}.$$

7. Then, for n modestly large, under the null hypothesis, W is approximately distributed as $\mathcal{N}(0, \text{Var}(W))$.

With the present data set, the null hypothesis is rejected by the Wilcoxon signed rank test at the 0.0409 level.

Below, we consider a class of tests based very much on the notions of Fisher's nonparametric test of the hypothesis that cross-fertilized and self-fertilized plants are drawn from the same distribution when it comes to the question of height. It would not be unreasonable, in fact, to note that Efron's bootstrap [5] is simply a sampling simulation approximation to the Fisher nonparametric test. Moreover, Fisher could have tried to come up with nonparametric tests with significance worked out combinatorially in more complicated situations. That he did not do so was probably due as much to his unwillingness to work out the messy algebra as to his faith in the ubiquity of the normal distribution. It was the brilliant contribution of Efron to note that we could come up with computer-intensive sampling approaches which would be asymptotically (as the number of resamplings went to infinity) equivalent to tests where exact significance levels would require highly complicated and (in the light of sampling using the high-speed computer) pointless symbolic manipulations. It should be noted that Fisher's example stood in plain view for 44 years before Efron could take what in retrospect appears to be an obvious computer age extension of Fisher's approach.

10.3   A Bootstrap Approximation to Fisher's Nonparametric Test

In our day, digital computing removes the necessity of going through such a combinatorial argument as that required to implement Fisher's nonparametric test for equality of differences. We can simply look at the differences one gets in 5,000 resamplings where we randomly shift from cross-fertilized to self-fertilized in each pair. When this was done we obtained the histogram in Figure 10.1.

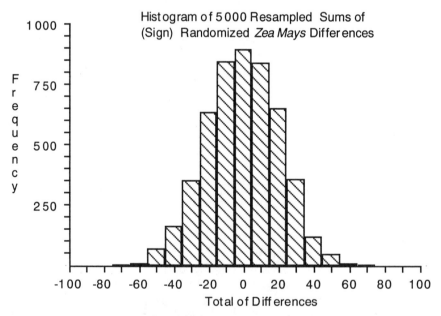

Figure 10.1. Resampled Sum of Differences with Random Sign Change.

In only 122 of these was the sum of differences greater than 39.25. Thus, our resampling test would reject the null hypothesis at the $2 \times 122/5000$

= .0488 level of significance. The significance level is off by a bit from the exact combinatorial answer, but not by much.

Now, it is a well-known fact that experimental investigators can have a tendency to present data to the statistician so as to promote the best chance for a significant argument in favor of significance of the data in support of an hypothesis. We note that the pairings in each pot might have been made in some way that this would be achieved. Better, then, to consider looking at differences in which the pairings are achieved randomly. Accordingly, we use the following resampling strategy (see Figure 10.2):

Resampling Test Based on Fisher's Nonparametric Test

1. Sample (with replacement) three of the crossed plants in Pot I and then sample (with replacement) three of the self-fertilized plants in Pot I.

2. Compute the differences between crossed and self-fertilized plants in each run where pairings are done randomly.

3. Sum the differences.

4. Carry out similar resamplings from each of the four pots.

5. Compute the sum of differences.

6. Repeat 5000 times.

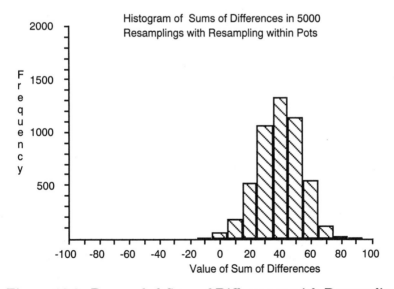

Figure 10.2. Resampled Sum of Differences with Resampling Within Pots.

Only 18 of the 5000 simulations gave a sum less than zero. The assumption of equality of stalk heights would be rejected at the $2 \times 18/5{,}000 = .007$ level. The mean of the difference distributions computed in this way is 39.48.

10.4   The Classical Randomization Test

Next, let us return to the analysis of Fisher, where he notes that the sum of the differences between cross-fertilized and self-fertilized is 39.25 inches. Let us consider an approach where we pool, for each pot, the crossed and self-fertilized plants. By resampling from such a set of pooled data, we can determine a distribution of summed differences in the situation where we disregard any information about cross-fertilized or self-fertilized. Our procedure for a *randomization test* is as follows:

Randomization Test Within Pots

1. Pool the six plants in Pot I.

2. Take a sample of size three (with replacement) from the pool and treat these as though they came from the cross-fertilized group.

3. Take a sample of size three (with replacement) from the pool and treat these as though they came from the self-fertilized group.

4. Difference the cross-fertilized and self-fertilized heights and sum the differences.

5. Carry out the same procedure in each of the rest of the pots.

6. Sum the 15 differences.

7. Repeat 5000 times.

8. Plot the histogram, which is *de facto* the distribution of the null hypothesis (see Figure 10.3).

9. Compare the difference observed in the actual test (39.25) with the percentiles of the randomization null hypothesis distribution.

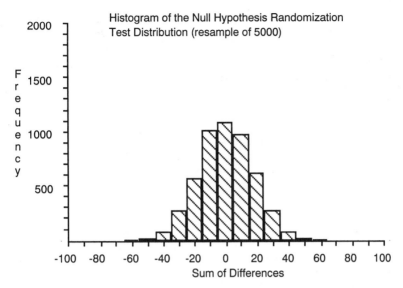

Figure 10.3. Null Hypothesis Distribution of Randomization Test.

Fifty out of the 5,000 "null distribution" observations of the randomization test were 39.25 or greater. Thus, using the randomization test above, we would reject the null hypothesis at the $2\times50/5,000 = .02$ level.

10.5   A Resampling-Based Sign Test

Let us suppose that one of the self-fertilized plants had been much taller. Suppose, for example, that the 12.75 value in Pot IV had been 22.75. Then, the t statistic drops to 1.721 and rejection of the null hypothesis would only be at the .11 level.

Suppose we disregard the sizes of the differences, relying completely on the signs of the differences. Then, we note that we would have, with the inflated value of 22.75 for the self-fertilized plant in Pot IV, three negative and twelve positive signs. The probability of having three negatives or three positives if the probability of any of the differences being positive (or negative) is $1/2$, is given by

$$2\sum_{j=0}^{3} \frac{15!}{j!(15-j)!} \left(\frac{1}{2}\right)^{15} = .035. \qquad (10.4)$$

Once again, there is the possibility that the ordering of the pairs might favor rejection of the null hypothesis. Let us consider a variation of the randomization within-pots approach, where we count the number of positive differences between cross-fertilized and self-fertilized plants in resampled pairs of size 15. The number of simulations is 5000. We note that we have no runs at all in which the number of positive differences is less than six. The number less than eight is only 9. If the heights of crossed and self-fertilized plants were distributed the same within pots, we should have around 2500 less than eight. The sign test, constructed as above, gives rejection of the null hypothesis of height equality very close to the .000 level.

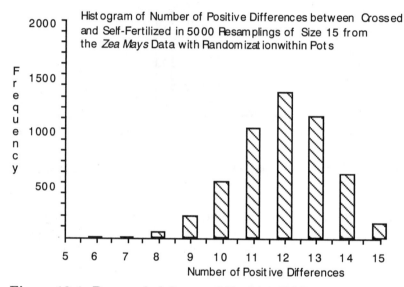

Figure 10.4. Resampled Sums of Positive Differences with Resampling Within Pots.

10.6   A Resampling Approach to Confidence Intervals

Fisher noted a powerful symmetry in the expression of the t-test. Let us consider the usual statement of the t-test:

$$P\left(-t_{.025} \leq \frac{\bar{x} - \mu}{s/\sqrt{n}} \leq t_{.025}\right) = .95. \qquad (10.5)$$

Now, from the standpoint of the observer, $\bar{x}$ is known. It is μ that is unknown. So Fisher rewrote the formula as

$$P(\bar{x} - t_{.025}s/\sqrt{n} \leq \mu \leq \bar{x} + t_{.025}s/\sqrt{n}) = .95. \qquad (10.6)$$

For the *Zea mays* data, the average difference $\bar{d}$ is given by 2.6167 and the standard deviation s_d by 4.7181. $n = 15$. The t value (with $n - 1 = 14$

degrees of freedom) is 2.145. So, then, the 95% confidence interval for $d = \mu_{\text{crossed}} - \mu_{\text{self}}$ is given by

$$2.6167 - 2.145 \times 4.7181/\sqrt{15} \leq d \leq 2.6167 + 2.145 \times 4.7181/\sqrt{15}$$

This tells us that 95% confidence interval for d is (.0036, 5.2298).

For a naive resampling strategy for obtaining a 95% confidence interval for d, we can look at the histogram in Figure 10.2. If we look at the 125th and 4875th resampled summed differences, we find 9.123 and 62.875, respectively. Dividing by the sample size (15), we obtain the naive resampled confidence interval as (.608, 4.192). We note that this interval is substantially shorter than that obtained using the t statistic. The mean of the sum of the differences is 39.48, giving, as the estimator for d, 2.632. The difference of the upper end of the resampled confidence interval and 2.632 is 1.560. The difference between 2.632 and the lower end of the confidence interval is 2.029. As we look at Figure 10.2, the slight skewing to the left of the summed differences is clear.

Actually, of course, the procedure whose histogram is shown in Figure 10.2 is not really quite a surrogate for Fisher's t-test. In Figure 10.2 we sampled randomly within pots, rather than letting the differences appear as delivered by Darwin to Galton. Figure 10.5 shows the histogram of 5000 sums of resamples (with replacement) of 15 differences using as the database the differences in Table 10.1.

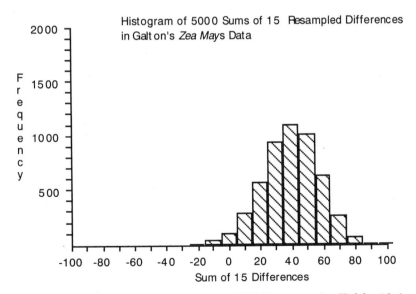

Figure 10.5. Resampled Sums of Differences in Table 10.1.

The 125th and 4875th smallest values were 4 and 72, respectively, with a mean of 39.506. Dividing by the sample size (15) we have the 95% confidence interval (.267, 4.800). The difference between the mean estimator

$39.506/15 = 2.634$ and 4.800 is 2.166, and the difference between $.267$ and 2.634 is 2.37. The histogram is less skewed to the left than the "within-pot" histogram of Figure 10.2.

The philosophical comparisons between the classical confidence intervals given by an argument based on the t-distribution and the various ones based on resampling strategies are nontrivial. In the resampling approaches, we are saying things like: "95% of the resampled average differences were between $.267$ and 4.800." In the case of the parametric t-based procedure, we are saying, "with probability 95%, the true value of d lies between $.0036$ and 5.2298." The two statements are not equivalent.

10.7   Resampling for Regression Model Selection

The problem of dealing with outliers is one that has produced thousands of papers in the scientific literature. When shall we remove, or discount, an observation that appears to be well outside the range of the rest of the database? The hard part really is deciding when an observation is truly atypical of the rest of the data. This is particularly a problem when the data is multidimensional. We need a procedure that is more or less automatic, which does not require us constantly to be attempting graphing (in high dimensions, we should talk rather of "pseudographing," since we will be looking at projections which usually fail to capture the essence of the data) the data and our model(s) of the systems that generated the data.

One attractive "rough-and-ready" procedure is iteratively reweighted least squares [3]. Let us suppose that we are attempting to model a response y by some function f of predictor variables $x_1, x_2, \ldots, x_p$. Then, according to the algorithm of iteratively reweighted least squares, we will use as criterion function:

$$Q = \sum_{i=1}^{n} (\hat{y}_i - y_i)^2 \frac{\nu + 1}{\nu + (r_i/s)^2}. \tag{10.7}$$

where the residual $r_i = \hat{y}_i - y_i$ is from the preceding iteration, and s^2 is the average of the n residuals from the preceding iteration. (The constant ν reflects our belief in just how taily the error distribution might be, à la, the degrees of freedom in a t-distribution. One fairly conservative value is $\nu = 3$.) In other words, iteratively reweighted least squares lowers the relative weighting on observations that deviate most seriously from the fitted model. One of the beauties of the algorithm is that it works with any number of predictor variables and with arbitrary predictor functions f. The algorithm generally converges without mishap. A disadvantage of the procedure might be that if we are in a situation where we need to use it, we probably will not be happy with using the usual $\mathcal{F}$ ratio tests.

For the purposes of demonstration, below we use a very simple one-dimensional example of Box and Draper [3]. We have 11 data points to which we wish to fit a straight line. The seventh seems a bit bizarre. So we fit it using an iteratively reweighted least squares approach with $\nu = 3$. The resulting estimator is $y_{irls} = -32.64 + 13.63x$. The weights for the convergent solution are shown in Table 10.2. Now the standard least squares estimator here is $y_{ls} = -33.02 + 13.83x$. We wish to decide which of the two predictors appears to be better, based on the absolute error of prediction of each estimator.

Table 10.2

x	y	y_{ls}	Weight	$|y_{ls} - y|$	y_{irls}	$|y_{irls} - y|$	Improv.
4.100	24.000	23.243	1.30	0.757	23.683	0.317	0.440
4.500	25.000	28.695	.98	3.695	29.215	4.215	-0.520
4.700	30.000	31.421	1.25	1.421	31.981	1.981	-0.560
4.800	32.000	32.784	1.31	0.784	33.364	1.364	-0.580
5.100	39.000	36.873	1.18	2.127	37.513	1.487	0.640
5.300	40.000	39.599	1.33	0.401	40.279	0.279	0.122
5.600	58.000	43.688	.44	14.312	44.428	13.572	0.740
5.800	48.000	46.414	1.24	1.586	47.194	0.806	0.780
6.200	48.000	51.866	.97	3.866	52.726	4.726	-0.860
6.500	54.000	55.955	1.18	1.955	56.875	2.875	-0.920
7.000	63.000	62.770	1.33	0.230	63.790	0.790	-0.560

Let us examine a histogram of the improvements in absolute difference between predicted and observed comparing iteratively reweighted least squares ($\nu = 3$) versus ordinary least squares results. Interestingly, a significant improvement seems not to have occurred as indicated from Figure 10.6.

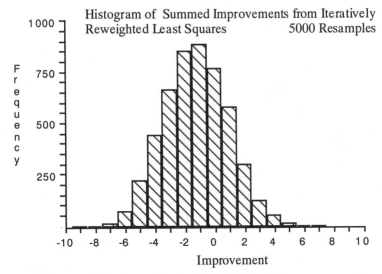

Figure 10.6. Iteratively Reweighted Improvements.

But what if we leave out the clearly bad pair (5.6, 58.0)? We show the results in Figure 10.7 with 5000 resampled sums of improvements leaving out the bad pair in the resampling. A significant improvement by using reiteratively reweighted least squares seems not to have occurred.

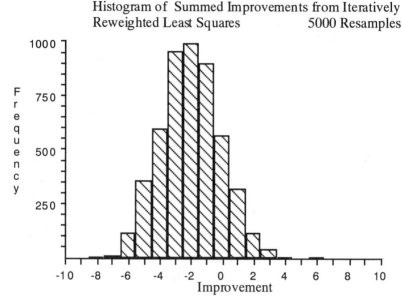

Figure 10.7. Iteratively Reweighted Improvements.

10.8   The Bootstrap as an All-Purpose Tool

It is seldom the case that a data analyst is called in at the experimental design stage of a study. Most of the time, we have data presented to us after completion of the study and are asked to make some sense of what the study shows. Dealing with such situations by parametric means requires a catalogue of techniques, each appropriate in some situations but disastrous in others. Picking and choosing when to use an $\mathcal{F}$ test or a rank procedure is part of the general arsenal of the statistician. And it requires a fair amount of at least modest-level mathematics to grasp what is going on.

For some years, some resamplers (see, e.g., [10]) have advocated dispensing with distribution theoretic considerations and simply using common sense to design a bootstrap approach to deal with the situation. Let us consider a few examples of how this might be done.

10.8.1   A Question from the Board of Education

There is frequently concern over the methodology used for improving students' reading skills. Particularly in suburban school districts, there is an attempt to improve upon the standard methodology. In this instance, the statistician is confronted with data from three different ways of improving the reading skills of fourth graders. In one school, there were three different classes, each taught by different methodologies. The first method, call it **A**, is the standard that has been used for five years. The second method, call it **B**, differs only from the standard one in that a different textbook is used. The third method, call it **C**, is markedly different from the first two. In **C**, there is intervention in the form of extensive instruction, in small student groups, by doctoral students in education from a nearby state university. Whereas **A** and **B** have essentially the same cost, method **C** would, absent the intervention for free by five doctoral students and their advisor, be very costly indeed.

At the end of the year, a standardized test is given to students from all three groups, and their scores are measured as departures from the scores of students in the past five years taking the same test. The scores are given in Table 10.3.

We have 30 tested students from **A**, 21 from **B**, and 24 from **C**. Of course, we can point out to the Board of Education that our task has not been made easier by the fact that there are three different teachers involved and the "teacher effect" is generally important. But that will not do us much good. The Board points out that it would have been practically impossible, under the limitations of size of the school, to have eliminated the teacher effect. We can, supposedly, take some comfort from the fact that the principal points out that the teachers, all similar in age and background, are uniformly "excellent." We are also assured that, in the interests of multicultural breadth, students are randomly mixed into classes from year to year. In any event, the data are as presented, and it is pointless to dwell too much on what might be done in the best of all possible worlds. Telling a client that "this job is just hopeless the way you have presented it to me" is not very good for keeping the consultancy going. Moreover, such an attitude is generally overly pessimistic.

One intuitive measure of the effectiveness of such programs is the mean score for each of the methods: 0.73, 2.4, and 3.64, respectively. The standard deviations are 2.81, 1.78, and 3.53, respectively.

The first thing we note is that the improvements, if any, are modest. The principal replies that such incrementalism is the way that improvements are made. She wishes to know whether or not the improvements for each of the two new methods, **B** and **C** are real or may simply be disregarded as due to chance. In Figure 10.8, we show a histogram of bootstrapped means using a resampling size of 10,000.

Table 10.3. Standardized Reading Scores.		
Group A	Group B	Group C
3.5	2.5	-1.4
4.8	2.5	4.9
-4.4	1.7	5.3
-3.2	3.2	1.4
1.9	5.1	5.3
0.0	3.0	0.4
1.3	5.4	4.8
3.6	-1.1	8.2
4.0	1.5	9.8
0.5	1.4	5.4
0.3	0.5	-3.6
-2.8	1.2	3.8
-0.4	3.8	3.0
-2.5	1.6	6.0
5.4	5.4	5.4
0.0	0.2	3.1
-3.0	2.9	4.9
3.1	3.5	2.6
3.6	3.8	1.5
-0.2	2.4	9.9
-1.1	-0.2	4.3
-3.9		1.4
2.1		-4.1
5.7		4.9
1.7		
3.0		
-2.0		
0.7		
0.4		
-0.3		

Now, the mindset of the bootstrapper is to consider the data to represent all reality, all possibilities that can ever occur. So, as a first step, we could compare means of samples (with replacement) of size 21 from **B** with means of samples (with replacement) of size 30 from the current pedagogy, **A**. In Figure 10.8 we show the histogram of differences between the two procedures using 10,000 resamplings. Here, denoting the resampled sample mean by μ, we note that the average performance of a resampled class using methodology **B** is greater than that of methodology **A** in over 91% of the runs. This gives us a bootstrapped "significance level" of .086, that is, the chances are only 8.6% that, if the methodologies were equally effective based on class average scores on the standardized test, that we would have seen a performance difference as large or larger than that which we have observed. Since Fisher, the standard for instituting change has been a rather arbitrary .05 or .01 level. We have seen that in SPC, we use an even more stringent .002 level. But significance levels should be adjusted to the reality of the situation. The cost of changing textbooks is marginal. The Board of Education might be well advised, therefore, to consider moving to methodology **B**.

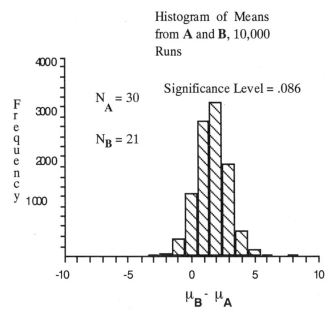

Figure 10.8. Means Histogram of B Versus A.

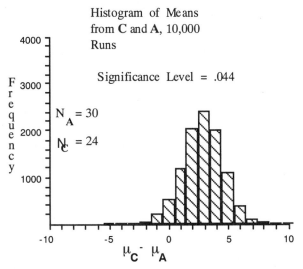

Figure 10.9. Means Histogram of C Versus A.

Next, let us investigate the resampling result when testing means of the costly procedure, **C**, versus that of the old standard, **A**. We demonstrate these results in Figure 10.9.

There seems little question as to the superiority of **C** to **A**. We see a bootstrap significance level of .044. On the other hand, we recall that **C** was an experimental, labor-intensive protocol that would be difficult to implement. Perhaps we should raise the question as to how much better it is than the cheap **B** protocol. We give the resampled comparisons of mean scores from **C** and **B** in Figure 10.10. If the two procedures had equal efficacy, then we would have observed a difference as great as we have observed 22.2% of the time.

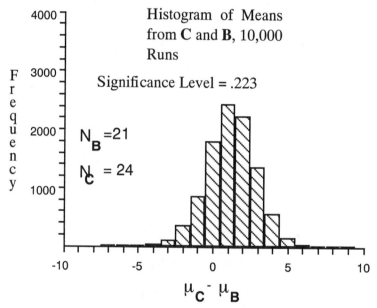

Figure 10.10. Means Histogram of C Versus B.

There may be another advantage for **B** over **C**. Let us carry out 10,000 resamplings of the difference between a class standard deviation from the **C** group with that from the **B** group. We note in Figure 10.11 that if there were no difference in the two, from a standard deviation standpoint, then we would have seen as large a difference as that observed in only 13% of the cases.

Our analysis does not give the Board of Education an unambiguous call. But it would appear that the cost of going to plan **B** (i.e., changing to the new textbook) may very well be the way to go.

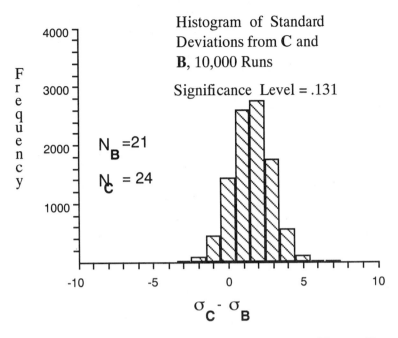

Figure 10.11. Standard Deviation Histogram of C Versus B.

10.8.2   Is This Forecasting Package Worth It?

One of the standard overhead costs for a stock brokerage house is forecasting software. These packages are generally pricey. The new senior partner of Medium Term Capital Management, Inc., is rather annoyed with the costly ACME package, which she believes performs poorly.

The package simply gives one week ahead forecasts for a popular index of small capitalization stocks. The senior partner is contemplating canceling the forecasting service. She thinks the forecasting is really no better than betting the index will not change during that week. She collects a list of the errors of 26 weeks of errors (expressed in percentage of the value of the index at the end of the week), and the errors of 26 weeks of percentage errors when the forecast is that the index will not change during the week. We show these in Table 10.4. Also, in Table 10.4, we show a column obtained in the following way. First, we have taken absolute values of all the errors, and then take the difference between the errors for the no-change forecast and that of the ACME forecast. We next carry out a bootstrap test where we take 10,000 resampled differences of size 26 and look at the means, as shown in Figure 10.12.

Table 10.4. Percentage Forecast Errors.			
Week	ACME Errors	"No Change" Errors	|NC Error|-|Acme Error|
1	-1.2	2.9	1.7
2	-0.2	2.8	2.5
3	0.2	-1.2	1.0
4	-0.5	0.2	-0.3
5	-0.7	-2.2	1.4
6	0.1	0.8	0.7
7	0.9	-3.6	2.7
8	-1.0	1.6	0.7
9	-1.1	6.6	5.5
10	1.2	1.3	0.1
11	0.3	1.6	1.2
12	2.0	2.1	0.1
13	-1.4	0.0	-1.4
14	0.1	-1.0	0.9
15	-0.6	-3.3	2.7
16	-0.5	-1.9	1.5
17	0.8	1.0	0.2
18	-0.4	2.9	2.5
19	-0.1	0.7	0.6
20	-0.2	1.3	1.1
21	0.5	-0.4	-0.1
22	0.3	1.5	1.2
23	0.4	0.6	0.2
24	0.0	0.0	0.0
25	0.2	0.9	0.8
26	0.6	1.7	1.1

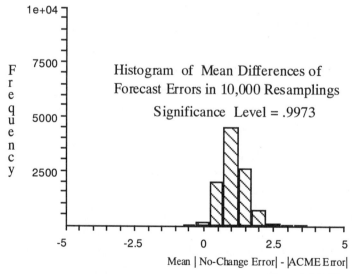

Figure 10.12. ACME Versus No-Change Forecast.

Our report to the senior partner is that the ACME package is a significant (.9973, bootstrap significance level) improvement over the "no-change" forecast. The ACME package indeed is a bit disappointing. Perhaps we could, with a bit of empirical time-series work, come up with a better forecast model. But ACME appears to be better than assuming the market will be static one week into the future.

10.8.3   When the Data Are Close to Normal

There are many cases where we can, due to central limit theory considerations, essentially treat the data as if it is close to Gaussian. When this is the case, will bootstrap methods be ruinous? The answer, as experienced nonparametricians might expect, is "Not necessarily. It all depends on the sparcity of the data. And for one dimension, nonparametrics can frequently work with apparently sparse data." As an example, let us consider a control chart example from Thompson and Koronacki ([11], pp. 77–87). Here, we have a situation where the basic process produces bolts of diameter 10 mm, but there are contaminations intermittent in time, so that the observable variable $Y(t)$ in any given lot is given by

$$Y(t) = \mathcal{N}(10, .01); \text{ probability } .98505$$
$$Y(t) = \mathcal{N}(10.4, .03); \text{ probability } .00995$$
$$Y(t) = \mathcal{N}(9.8, .09); \text{ probability } .00495$$
$$Y(t) = \mathcal{N}(10.2, .11); \text{ probability } .00005.$$

Now in Table 10.5 we show the results of a sampling of 90 lots, each of size five. It turns out lot 28 and 79 come from $\mathcal{N}(10.4, .03)$. Lot 40 comes from $\mathcal{N}(9.8, .09)$. And all the other lots come from the uncontaminated $\mathcal{N}(10, .01)$.

Table 10.5

Lot	x_1	x_2	x_3	x_4	x_5	$\bar{x}$	s	R	s^2
1	9.927	9.920	10.170	9.976	9.899	9.978	0.111	0.271	0.012
2	9.862	10.003	9.829	9.824	10.077	9.919	0.114	0.253	0.013
3	10.061	10.089	9.950	9.929	9.935	9.993	0.076	0.160	0.006
4	9.820	10.066	10.062	9.897	10.013	9.972	0.109	0.246	0.012
5	9.737	9.937	9.928	10.144	9.965	9.942	0.145	0.406	0.021
6	9.876	9.957	9.845	9.913	9.941	9.906	0.046	0.112	0.002
7	9.898	9.959	9.924	9.989	9.987	9.951	0.040	0.092	0.002
8	10.001	10.050	10.263	9.982	10.076	10.074	0.112	0.281	0.013
9	9.928	10.234	9.832	10.027	10.121	10.028	0.158	0.402	0.025
10	9.896	9.994	10.009	9.835	10.162	9.979	0.125	0.327	0.016
11	10.011	10.011	10.090	10.095	10.120	10.065	0.051	0.108	0.003
12	9.983 9	.974	10.071	10.099	9.992	10.024	0.057	0.125	0.003
13	10.127	9.935	9.979	10.014	9.876	9.986	0.094	0.251	0.009
14	10.025	9.890	10.002	9.999	9.937	9.971	0.056	0.136	0.003
15	9.953	10.000	10.141	10.130	10.154	10.076	0.092	0.201	0.009
16	10.007	10.005	9.883	9.941	9.990	9.965	0.053	0.124	0.003
17	10.062	10.005	10.070	10.270	10.071	10.096	0.101	0.266	0.010
18	10.168	10.045	10.140	9.918	9.789	10.012	0.158	0.379	0.025
19	9.986	10.041	9.998	9.992	9.961	9.996	0.029	0.080	0.001
20	9.786	10.145	10.012	10.110	9.819	9.974	0.165	0.359	0.027
21	9.957	9.984	10.273	10.142	10.190	10.109	0.135	0.316	0.018
22	9.965	10.011	9.810	10.057	9.737	9.916	0.137	0.321	0.019
23	9.989	10.063	10.148	9.826	10.041	10.013	0.119	0.322	0.014
24	9.983	9.974	9.883	10.153	10.092	10.017	0.106	0.270	0.011
25	10.063	10.075	9.988	10.071	10.096	10.059	0.041	0.108	0.002
26	9.767	9.994	9.935	10.114	9.964	9.955	0.125	0.347	0.016
27	9.933	9.974	10.026	9.937	10.165	10.007	0.096	0.232	0.009
28	10.227	10.517	10.583	10.501	10.293	10.424	0.154	0.356	0.024
29	10.022	9.986	10.152	9.922	10.101	10.034	0.091	0.124	0.002
30	9.845	9.901	10.020	9.751	10.088	9.921	0.135	0.337	0.018

Lot	x_1	x_2	x_3	x_4	x_5	$\bar{x}$	s	R	s^2
31	9.956	9.921	10.132	10.016	10.109	10.027	0.092	0.212	0.009
32	9.876	10.114	9.938	10.195	10.010	10.027	0.129	0.318	0.017
33	9.932	9.856	10.085	10.207	10.146	10.045	0.147	0.352	0.022
34	10.016	9.990	10.106	10.039	9.948	10.020	0.059	0.158	0.003
35	9.927	10.066	10.038	9.896	9.871	9.960	0.087	0.195	0.008
36	9.952	10.056	9.948	9.802	9.947	9.941	0.090	0.254	0.008
37	9.941	9.964	9.943	10.085	10.049	9.996	0.066	0.144	0.004
38	10.010	9.841	10.031	9.975	9.880	9.947	0.083	0.190	0.007
39	9.848	9.944	9.828	9.834	10.091	9.909	0.112	0.262	0.013
40	10.002	9.452	9.921	9.602	9.995	9.794	0.252	0.550	0.064
41	10.031	10.061	9.943	9.997	9.952	9.997	0.050	0.118	0.003
42	9.990	9.972	10.068	9.930	10.113	10.015	0.074	0.183	0.006
43	9.995	10.056	10.061	10.016	10.044	10.034	0.028	0.066	0.001
44	9.980	10.094	9.988	9.961	10.140	10.033	0.079	0.179	0.006
45	10.058	9.979	9.917	9.881	9.966	9.960	0.067	0.176	0.004
46	10.006	10.221	9.841	10.115	9.964	10.029	0.145	0.380	0.021
47	10.132	9.920	10.094	9.935	9.975	10.011	0.096	0.212	0.009
48	10.012	10.043	9.932	10.072	9.892	9.990	0.076	0.179	0.006
49	10.097	9.894	10.101	9.959	10.040	10.018	0.090	0.207	0.008
50	10.007	9.789	10.015	9.941	10.013	9.953	0.097	0.226	0.009
51	9.967	9.947	10.037	9.824	9.938	9.943	0.077	0.213	0.006
52	9.981	10.053	9.762	9.920	10.107	9.965	0.134	0.346	0.018
53	9.841	9.926	9.892	10.152	9.965	9.955	0.119	0.311	0.014
54	9.992	9.924	9.972	9.755	9.925	9.914	0.093	0.236	0.009
55	9.908	9.894	10.043	9.903	9.842	9.918	0.075	0.201	0.006
56	10.011	9.967	10.204	9.939	10.077	10.040	0.106	0.265	0.011
57	10.064	10.036	9.733	9.985	9.972	9.958	0.131	0.330	0.017
58	9.891	10.055	10.235	10.064	10.092	10.067	0.122	0.345	0.015
59	9.869	9.934	10.216	9.962	10.012	9.999	0.132	0.346	0.017
60	10.016	9.996	10.095	10.029	10.080	10.043	0.042	0.099	0.002
61	10.008	10.157	9.988	9.926	10.008	10.017	0.085	0.231	0.007
62	10.100	9.853	10.067	9.739	10.092	9.970	0.165	0.361	0.027
63	9.904	9.848	9.949	9.929	9.904	9.907	0.038	0.101	0.001
64	9.979	10.008	9.963	10.132	9.924	10.001	0.079	0.208	0.006
65	9.982	9.963	10.061	9.970	9.937	9.983	0.047	0.124	0.002
66	10.028	10.079	9.970	10.087	10.094	10.052	0.052	0.123	0.003
67	9.995	10.029	9.991	10.232	10.189	10.087	0.115	0.241	0.013
68	9.936	10.022	9.940	10.248	9.948	10.019	0.133	0.312	0.018
69	10.014	10.070	9.890	10.137	9.901	10.002	0.107	0.247	0.011
70	10.005	10.044	10.016	10.188	10.116	10.074	0.077	0.183	0.006
71	10.116	10.028	10.152	10.047	10.040	10.077	0.054	0.124	0.003
72	9.934	10.025	10.129	10.054	10.124	10.053	0.080	0.195	0.006
73	9.972	9.855	9.931	9.785	9.846	9.878	0.074	0.187	0.005
74	10.014	10.000	9.978	10.133	10.100	10.045	0.068	0.155	0.005
75	10.093	9.994	10.090	10.079	9.998	10.051	0.050	0.098	0.003
76	9.927	9.832	9.806	10.042	9.914	9.904	0.093	0.236	0.009
77	10.177	9.884	10.070	9.980	10.089	10.040	0.112	0.293	0.013
78	9.825	10.106	9.959	9.901	9.964	9.951	0.103	0.281	0.011
79	10.333	10.280	10.509	10.631	10.444	10.439	0.140	0.204	0.006
80	9.972	10.116	10.084	10.059	9.914	10.029	0.084	0.202	0.007
81	10.059	9.992	9.981	9.800	9.950	9.956	0.096	0.259	0.009
82	9.832	10.075	10.111	9.954	9.946	9.984	0.112	0.279	0.012
83	9.958	9.884	9.986	10.008	10.113	9.990	0.083	0.229	0.007
84	10.087	9.994	9.915	10.023	9.883	9.980	0.082	0.205	0.007
85	10.232	9.966	9.991	10.021	9.965	10.035	0.112	0.324	0.014
86	10.066	9.948	9.769	10.102	9.932	9.963	0.131	0.333	0.017
87	10.041	10.044	10.091	10.031	9.958	10.033	0.048	0.133	0.002
88	9.868	9.955	9.769	10.023	9.921	9.907	0.096	0.254	0.009
89	10.084	10.018	9.941	10.052	10.026	10.024	0.053	0.143	0.003
90	10.063	10.055	10.104	10.080	10.064	10.073	0.019	0.049	0.001

The standard SPC procedure is to compute the pooled mean of the sample means (10.0044 in this case), the pooled mean of the sample standard deviations (.0944, in this case) and use these values to obtain a control chart in which a lot that actually comes from $\mathcal{N}(10, .01)$ would be rejected with probability only .002 (to minimize the chance of false alarms). The resulting chart is shown in Figure 10.13.

We see that here the control chart easily identifies the three contaminated lots. What would the situation be for a bootstrap procedure? Let us suppose we use a bootstrap confidence interval approach. We first estimate the pooled mean (10.0044). Then for each lot, we construct a histogram of the bootstrapped means and obtain the (.001, .999) confidence interval. If the interval contains the value 10.0044, then we declare it to be in control. Otherwise, we declare it to be out of control. We note that with a data set of five there are $5^5 = 3125$ possibilities. So the granularity does not seem to be an overwhelming problem. Let us try the procedure on lot 40, which we know is not in control. In Figure 10.14 we show the resulting histogram of bootstrapped means. The 99.8% confidence interval is (9.4820, 9.9964). Since this interval does not include the pooled mean for all the lots (10.0044), we reject the hypothesis that lot 40 is an in-control lot. Similarly, it is easy to see that we also reject as in-control the other two bad lots. Unfortunately, we also reject a number of lots which are, in fact, in control. These include lots 6, 7, 54, 63, and 73. Thus, using the bootstrap has caused a number of false alarms, which, as a practical matter, discourages workers from using SPC at all.

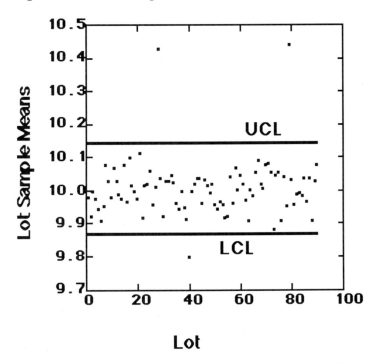

Figure 10.13. Mean Control Chart for 90 Lots of Bolts.

There are ways to make things better. Of course, these will require us to make some assumptions (we recall the nail soup example in Section 2.3). For example, let us suppose we make the assumption that all the lots come from the in-control distribution. Then we can take the overall pooled mean of sample means (10.044) and record all the means from all the lots as differenced from 10.044.

$$(diff)_j = \overline{X}_j - 10.0044.$$

It is an easy matter then to use bootstrapping to find a 99.8% confidence interval about the mean of sample means in which the population mean should lie. But, recalling that

$$\text{Var}(\overline{\overline{X}}) = \text{Var}(\overline{X_j})/90,$$

where j is simply one of the 90 lots, we realize that, under the assumptions given, the standard deviation of the lot means, should be taken to be $\sqrt{90}$ times that of the mean of lot means. So to find a confidence interval about 10.0044 in which a lot mean would be expected to fall with 99.8% chance, if it is truly from the in-control population, we construct a histogram of the bootstrapped differences multiplied by $\sqrt{90}$ with 10.0044 added. We show this in Figure 10.15.

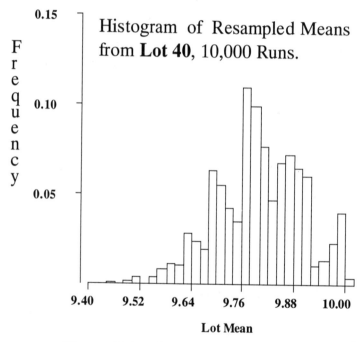

Figure 10.14. Bootstrapped Means for Lot 40.

The 99.8% confidence interval (a sort of bootstrapped mean control chart), is given by (9.7695,10.289). In Figure 10.16, we note that (contaminated) lots 28 and 79 both are out of control. But (contaminated) lot 40 is not recognized as being out of control.

We note that none of the uncontaminated lots is identified as out of control. Consequently, in this case, the bootstrap control chart worked satisfactorily. By using the deviation of lot means from grand mean, we have, however, made an assumption that very frequently gives an inflated measure of the in-control process variability and causes us to construct confidence intervals that are too wide, thus accepting too many bad lots as being in control.

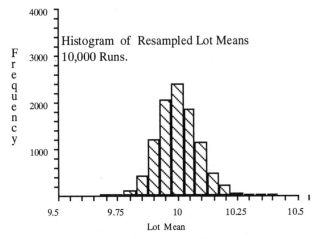

Figure 10.15. Bootstrapped Lot Means.

There are still other bootstrap strategies which we might employ, but we shall not go into such robustification techniques here. (The interested reader is referred to Thompson and Koronacki [11].) The fact is that we do pay a price in the SPC setting for bootstrapping instead of using the classical Gaussian-based Shewhart control chart when the assumption of Gaussianity is a reasonable one. Let it be understood, however, that computational complexity is hardly a justification these days for not using a resampling algorithm. It is an easy matter to build hand held calculators which will give bootstrapped results for a predetermined protocol in a very short time.

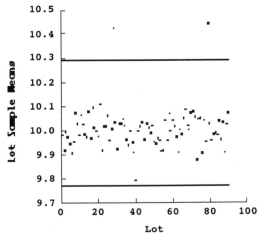

Figure 10.16. Bootstrapped Mean Control Chart.

10.8.4   A Bootstrapped Forecast

Table 10.6. Stock Price Ratios.			
Week $= n$	$S(n)$	$S(n{+}1)$	$R(n) = S(n+1)/S(n)$
1	44.125	42.500	0.963
2	42.500	45.375	1.068
3	45.375	47.625	1.050
4	47.625	50.250	1.055
5	50.250	50.250	1.000
6	50.250	48.875	0.973
7	48.875	49.125	1.005
8	49.125	45.875	0.934
9	45.875	46.750	1.019
10	46.750	45.125	0.965
11	45.125	47.875	1.061
12	47.875	49.250	1.029
13	49.250	51.875	1.053
14	51.875	53.000	1.022
15	53.000	51.875	0.979
16	51.875	47.750	0.920
17	47.750	48.125	1.008
18	48.125	50.125	1.042
19	50.125	48.875	0.975
20	48.875	47.875	0.980
21	47.875	46.125	0.963
22	46.125	48.125	1.043
23	48.125	49.125	1.021
24	49.125	49.000	0.997
25	49.000	51.250	1.046
26	51.250	50.875	0.993
27	50.875		

Possibly one of the difficulties with the bootstrap is that it inspires all sorts of model-free "great ideas" which can lead to ill-conceived algorithms in a twinkling. Let us consider one possible such idea. Since bootstrapping is increasingly replacing classical statistical courses in schools of business, let us take such an example in an equity setting. A broker is preparing an analysis for guiding her clients in the risks associated with a certain stock. To carry out this, she takes closing prices at the end of trade on Wednesdays for 27 consecutive weeks. She notes the ratio of the value of the stock on week $n + 1$ divided by that for week n, denoting the ratio as $R(n)$. A one-week interval is chosen in order to help with the assumption that the ratios from week to week are roughly stochastically independent. So, in Table 10.6, we show such stock values and such ratios: The results are interesting. Eleven of the ratios are less than 1; one is equal to 1; and fourteen are greater than 1. So far, so good. But now the broker wants

to go further. She decides to treat the historical ratios as representative of
what will happen in the future. So, in order to gain a profile of what is
likely to happen to an investment in the stock six months from now, she
bootstraps from the ratios samples of size 26. She then multiplies these
ratios together to come up with a sampled value for the ratio of the stock's
value in six months compared to what it is today. Having done this 2000
times, she constructs the histogram shown in Figure 10.17.

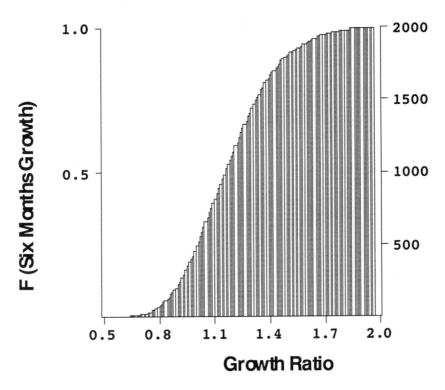

Figure 10.17. Bootstrapped Forecasts.

Based on that histogram, she tells clients that, based on the performance
of the last six months, the median growth rate of the stock over the next
six months will be 16% (annualized 32%) and the mean growth rate over
six months is 18% (annualized 36%). The probability of a loss greater than
30% of investment over six months is only 1%. Naturally, our broker can
make longer extrapolations. What about five years? No problem, we could
simply take samples with replacement of size 260 weeks. The programming
is trivial. But what of the assumptions? Clearly model-free extrapolation
over intervals of nontrivial length outside the database is risky business.

10.9   Empirical Likelihood: An Alternative to Resampling

The quest for getting by without assumptions about the distribution underlying a data set is ongoing. Owen [8] has proposed a strategy based, in large measure, on the multinomial likelihood function where the baseline (as in the bootstrap) is the Dirac-comb density function, which assumes that the underlying distribution is that obtained when one puts a Dirac mass at each one of the data points. Following Owen [8], let us assume we have a database of n points $X_1, X_2, \ldots, X_n$. The data points may be, essentially, of any Euclidean dimension, say d. Then, consider the empirical distribution function

$$F_n = \frac{1}{n} \sum_{i=1}^{n} H_{X_i}, \qquad (10.8)$$

where H_X is a distribution taking the value X with probability 1. Thus, $F(A)$ is the number of observations in A. Let F be any distribution on R^d. Let $L(F) = \prod_{i=1}^{n} F\{X_i\}$. Let us assume that in the data set $X_1, X_2, \ldots, X_n$ we have the k distinct values $Y_1, Y_2, \ldots, Y_k$. Let $p_i = F\{Y_i\}$, with $n_i \geq 1$. Then, using the fact that $\log(1 + z) \leq z$ and $\sum_i p_i \leq 1$, we have that

$$
\begin{aligned}
\log\left(\frac{L(F)}{L(F_n)}\right) &= \sum_{i=1}^{k} n_i \log\left(\frac{n p_i}{n_i}\right) \\
&\leq \sum_{i=1}^{k} n_i \left(\frac{n p_i}{n_i} - 1\right) \\
&\leq 0.
\end{aligned}
$$

Thus $L(F) \leq L(F_n)$; i.e., F_n is the nonparametric maximum likelihood estimator for the underlying distribution. The following rather surprising result holds:

Empirical Likelihood Theorem. Let $X, X_1, X_2, \ldots, X_n$ be independent and identically distributed random vectors in R^d, with $E(X) = \mu_0$, and $\mathrm{Var}(X)$ finite and of rank $q > 0$. For positive $R < 1$, let

$$C_{r,n} = \left[\int x \, dF(x) \mid R(F) \geq r, F(\{X_1, X_2, \ldots, X_n\}) = 1\right].$$

Then $C_{r,n}$ is a convex set and

$$\lim_{n \to \infty} P(\mu_0 \in C_{r,n}) = P(\chi_q^2 \leq -2 \log r).$$

Furthermore, if $E(\| X \|^4) < \infty$, then

$$|P(\mu_0 \in C_{r,n}) - P(\chi_q^2 \leq -2 \log r)| = O(n^{-1/2}).$$

We may use the *profile empirical likelihood ratio function* for computation purposes.

$$\mathcal{R}(\theta) = \sup\{R(F)|T(F) = \theta, F(\{X_1, X_2, \ldots, X_n\}) = 1\}. \qquad (10.9)$$

Confidence regions are of the form $\{\theta|\mathcal{R}(\theta) \geq r\}$. Tests of the null hypothesis that $\theta = \theta_0$ reject when $-2\log\mathcal{R}(\theta_0)$ is smaller than an appropriate critical χ^2 value.

When we are seeking a confidence region of the mean, we have

$$\mathcal{R}(\mu) = \max\left(\prod_{i=1}^{k}\left(\frac{np_i}{n_i}\right)^{n_i} \mid 0 \leq p_i, \sum_{i=1}^{k} p_i = 1\right). \qquad (10.10)$$

Now, for any give value of μ, we wish to find nonnegative weights $\{w_i\}$, which maximize $\prod_{i=1}^{n} w_i$ subject to $\sum_{i=1}^{n} w_i = 1$ and $\sum_{i=1}^{n} w_i X_i = \mu$. At **http://playfair.stanford.edu/reports/owen/el.S**, Owen gives an S program for carrying out these computations.

Let us now compare for a data set of size 20 from a normal distribution with mean 5 and standard deviation 2 the confidence intervals obtained using the classical t_{19} interval, that obtained by Owen's empirical likelihood, and that using Efron's bootstrap. The data set is (5.039, 3.167, 4.415, 6.469, 5.075, 8.664, 7.709, 3.893, 6.796, 4.118, 5.468, 0.612, 4.206, 5.576, 7.778, 4.527, 6.723, 4.793, 3.267, 5.527).

Now the sample mean $\bar{x}$ is given by 5.191, the sample standard deviation by 1.861, and the t_{19} value for a two-tailed 95% confidence interval is 1.729. Accordingly, the classical 95% confidence interval is given by $(\bar{x} - 1.729s/\sqrt{19}, \bar{x} + 1.729s/\sqrt{19})$ or (4.472,5.910).

In Figure 10.18 we show a cumulative histogram of 2000 resampled means of size 20. Going from the 50th largest to the 1950th largest gives a bootstrap 95% confidence interval of (4.392, 5.963).

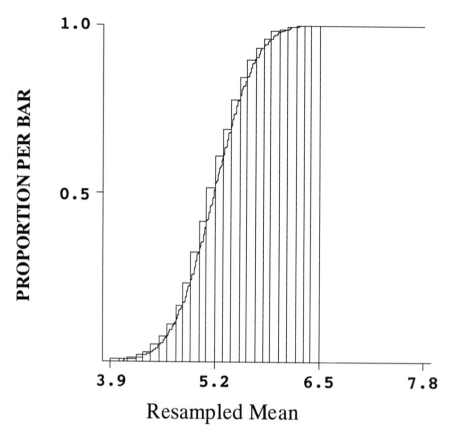

Figure 10.18. Bootstrap Means.

In Figure 10.19 we plot the logarithm of the empirical likelihood ratio function for varying values of μ. For this example, where the data is one-dimensional, the appropriate rank q for the χ^2 test is 1. Thus for the 95% critical χ^2 value, we have 3.841. That means that we should look for the interval consistent with $-3.841/2 = -1.920$. This gives us the 95% empirical likelihood confidence interval of $(4.314, 5.987)$.

Now, as one would have suspected, the confidence interval obtained when using the information that the data is really normal is smaller than that from either the bootstrap or the empirical likelihood. Naturally, if the data had come from a density with heavy tails, then this t interval could have been deceptively narrow (of course, we know that the t confidence interval has natural robustness properties if the tails are not too heavy). The χ^2 result is asymptotic, of course, and we might expect that the bootstrap would be more robust to departures from normality than the empirical likelihood function. It turns out that empirical likelihood is robust to heavy tails, but is confused by skewness.

We have not given examples where the empirical likelihood shines, that is, in problems of dimension greater than one. Empirical likelihood can be used to produce confidence sets for data of higher dimensions. In higher dimensions there appears at present to be a great deal of difficulty in producing confidence sets from the bootstrap.

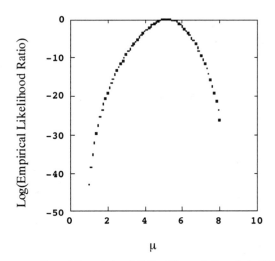

Figure 10.19. Log(Empirical Likelihood Ratio) Versus μ.

10.10   The Spectre at the Feast

10.10.1   The Good News

Around 20 years before Efron's bootstrap came on the scene, John Tukey [12] gave us the jacknife. Both the bootstrap and the jacknife achieve roughly the same goal: obtaining estimates of variability in situations where normality may not be a good assumption. Both have a nominalist, model-free, flavor. But, although Tukey's jacknife has been and continues to be put to good use, it never created the kind of boundless enthusiasm we have seen with Efron's bootstrap.

We have already noted that there were procedures before the bootstrap which were very like it. We have mentioned Fisher's permutation test [6]. Wilkinson [14] notes the presence of tests precursive to the bootstrap in the psychometric literature (e.g., Block [2]), and even in the theoretical statistical literature (e.g., Dwass [4]). None of these really "caught fire."

If ever there were evidence that the late Thomas Kuhn [7] had a point when he noted that there is a strong sociological component to the reasons the movements in science become established, surely the bootstrap is such evidence. It appeared at the onset of the personal computer. It appeared in perhaps the most innovative department of statistics in the world (Stanford, close by another great department, that at Berkeley), and it received the enthusiastic support of the members of that statistical community. It came at just the right time, in the sense that computing was, by 1979, essentially free. And it was sold effectively. At any rate, it has become a part of the statistical orthodoxy. Today, one might well get a paper published that uses bootstrapping, which might not get published if one uses, say, the jacknife. A drug will be accepted by the FDA as effective if bootstrapping techniques have been used to support the conjecture whereas, 15 years ago, it would not have been.

10.10.2   The Bad News

In his now classical *Exploratory Data Analysis* [13], Tukey called for what amounts to a revolution in statistical thinking. (Interestingly, much of Tukey's EDA work was carried out, not at his Princeton home base, but in his summers and leaves at Stanford where his bold ideas had great influence.) The old notions of tests and strategies based on an involved treatment of distributions was to be replaced by the intuition of the human observer viewing a battery of graphs. Models, distributional and otherwise, were largely to be swept away. This kind of Gestaltic nominalism was particularly significant coming as it did from one of the best theoretical mathematicians ever seen in the statistical sciences. Tukey's call to arms was not from a "know-nothing" but from one who had seen a whole gamut of distributional models and observed the damage that true believers in, say, universal Gaussianity, must experience, sooner or later.

Similarly, the bootstrap was also presented by a highly skilled mathematical scientist. Just as with EDA, or the jacknife, the bootstrap is supposed to help us deal with the ubiquity of non-Gaussian situations. And, to a very real extent, EDA and the jacknife and the bootstrap do help us cope with situations where the data are not Gaussian (normal).

A major issue in our decision is whether we need to model, somehow, the underlying distribution(s) of the system(s) under consideration in such simple matters as interpolation versus extrapolation. Most of us are familiar with interpolating in, say, tables of the normal distribution. If we are looking for the cdf value of standardized $\Phi(1.2134)$, say, generally we might be willing to use linear interpolation from the values for 1.21 and 1.22. What happens, however, when we are confronted with finding $\Phi(7.12)$, whose argument is well outside the range of most Gaussian tables? Most of us would find it prudent to find a very good approximation formula or even to use a fine mesh quadrature. It would be unlikely that we would try and extrapolate past $\Phi(4.00)$, the place where most normal tables stop, to, say, $\Phi(4.50)$.

As we have indicated in Section 10.4, in time-series, model-free techniques are unlikely to be advisable for extrapolation purposes. Giving up on the notion that one is looking at a Gaussian process causes one to lose the ability to replace ensemble averages with averages across time. And, not too surprisingly, for time-series problems the bootstrap has not proved very valuable. (But then, for market forecasting purposes, say, one can argue that, so far, nothing works very well for very long.)

If it were only in the stochastic process situation that we experienced some difficulty, that would be one thing. But, the reality is that one can get burnt for much simpler problems if one tries to use model-free techniques such as the bootstrap. For example, let us consider the problem of high dimensionality. Suppose that we have a random sample of size 100 over the unit interval and that the unit interval has been divided into a mesh of size .1. This gives us an expectation of 10 samples per tiling interval. Next, suppose that we have the same sample but now included in a hypercube of dimension four. Then we have an expectation of 100 hypercube tiles per observation. And so on. Essentially, for other than low-dimensional problems, even looking inside a data set involves extrapolation, for the data points will frequently be relatively few on a per hypervolume basis in a problem of high dimensionality. At the minimum, Cartesian tiling (hypercubes) will have to be replaced by a data-based nearest-neighbor tiling (as we discussed earlier with SIMDAT).

In the early 1980s, I received a request from a senior Army scientist, Dr. Malcolm Taylor, asking for a nonparametric density estimation—based resampler. The problem involved sampling from some 400 data points in four dimensions, which represented the results of penetrations of tank armor. My response was that he might look at the bootstrap. He had already done that, and in the way that the simulations were being used (tank design), some bizarre results were being obtained (e.g., the identification of "well-protected" spaces in the tank which were not, in fact, well protected). We have already shown in Chapter 5 how it is easy to see how such things can happen. A sample of size 100 of firings at a bull's-eye of radius 5 centimeters will, if the distribution of the shots is circular normal with mean the center of the bullseye and deviation one meter with a probability in excess of .88, have none of the shots hitting in the bull's-eye. So any Dirac-comb resampling procedure will tell us the bull's-eye is a safe place, if we get a base sample (as we probably will) with no shots in the bullseye. Again, we refer the reader to SIMDAT in Chapter 5 when dealing with continuous data in such a way that one is looking at other than estimating the equivalents of very low moments.

The situation is, of course, very much changed when we have a high-dimensional problem with some information about the generating distribution. Doing nonparametric data analysis in, say, eight space, with a sample of size 100, is simply very dubious. In the future as we move to data in higher dimensions, we will probably be well advised to follow the policy of first finding local regions of high density and then trying to model the stochastic mechanism approximating behavior in this region. Such a procedure is addressed in Chapter 11. However, the way most data are still analyzed is to look at low-order moments in problems of very low dimensionality, typically one. Here, Efron's bootstrap shines.

Problems

10.1. An IQ test is administered to students from two high schools in the same city (Table 10.7). Using a bootstrap procedure, comment upon the conjecture that both groups **A** and **B** have the same underlying IQ.

Table 10.7. Intelligence Quotient Scores.	
Group **A**	Group **B**
116.7	112.3
98.0	120.2
117.3	120.6
97.2	101.1
119.3	85.9
73.4	90.7
110.4	98.7
88.4	125.3
123.8	84.8
74.3	103.2
144.9	117.2
97.6	121.7
66.7	100.0
114.1	101.1
142.7	128.9
87.1	90.6
109.9	92.8
77.8	113.4
74.9	143.5
77.8	120.1
91.1	100.0
86.3	103.2
119.2	112.8
104.5	125.2
95.1	127.3
106.9	127.9
84.6	147.9
99.3	
96.9	
77.6	

10.2. For the two schools selected, the conjecture is made that there is a significant correlation between IQ and family income (measured in thousands of dollars). Use the data in Table 10.8 to obtain a bootstrap procedure for testing the conjecture.

Table 10.8. Intelligence Quotient Scores.			
GroupA	Income	GroupB	Income
116.7	15.1	112.3	21.4
98.0	20.3	120.2	66.2
117.3	25.7	120.6	45.1
97.2	56.3	101.1	23.1
119.3	45.2	85.9	19.1
73.4	70.2	90.7	22.1
110.4	19.1	98.7	21.1
88.4	14.2	125.3	45.2
123.8	72.4	84.8	11.1
74.3	14.2	103.2	74.1
144.9	97.3	117.2	44.1
97.6	36.0	121.7	97.2
66.7	13.2	100.0	23.1
114.1	36.1	101.1	19.3
142.7	19.1	128.9	35.6
87.1	44.7	90.6	22.1
109.9	55.1	92.8	13.1
77.8	72.1	113.4	23.8
74.9	15.1	143.5	101.3
77.8	13.9	120.1	87.1
91.1	19.1	100.0	44.4
86.3	56.2	103.2	28.1
119.2	34.1	112.8	10.5
104.5	45.1	125.2	36.7
95.1	24.8	127.3	12.3
106.9	16.2	127.9	28.1
84.6	23.1	147.9	15.2
99.3	18.1		
96.9	39.9		
77.6	15.2		

10.3. Using the data in Table 10.6, obtain bootstrapped forecasts for the stock performance two years forward.

10.4. Let us assume that we wish to obtain a bootstrapped mean control chart based on the data in Table 10.5. However, see what the situation will be if, as frequently happens, lot sizes of less than five are used. Namely, obtain a mean control chart based on mean data from the first three observations in each lot.

10.5. Again, using the data in Table 10.5, obtain a bootstrapped 99.8% confidence interval for σ based on the lot estimates s.

10.6. The senior partner changes her mind and decides that really all she cares about is whether the error difference between ACME and no-change is positive or negative. In other words, "a miss is as good as a mile." Comment on whether this should change the decision about continuing to buy the ACME forecasting package.

10.7. Now, even before estimating θ_A and θ_B in Problem 9.1, we see that adding on the counseling appears to have good effect. However, it is possible that this appearance is simply the result of randomness. And the counseling is not without cost. Orthodox Bayesians are generally reluctant to construct significance tests. But construct a resampling test to determine if it is realistic to assume that the counseling is of no benefit.

References

[1] Baggerly, K.A. (1998). "Empirical likelihood as a goodness-of-fit measure," *Biometrika*, **85**, 535–547.

[2] Block, J. (1960). "On the number of significant findings to be expected by chance," *Psychometrika*, **25**, 158–168.

[3], Box, G.E.P. and Draper, N.R. (1987). *Empirical Model-Building and Response Surfaces*. New York: John Wiley & Sons, 84–90.

[4] Dwass, M. (1957). "Modified randomization sets for nonparametric hypotheses," *Ann. Math. Statist.*, **29**, 181–187.

[5] Efron, B. (1979). "Bootstrap methods: another look at the jacknife," *Ann. Statist.*, **8**, 1–26.

[6] Fisher, R.A. (1935). *The Design of Experiments*. Edinburgh: Oliver & Boyd, 29–49.

[7] Kuhn, T. (1970). *The Structure of Scientific Revolutions*. Chicago: University of Chicago Press.

[8] Owen, A.B. (1988) "Empirical likelihood ratio confidence intervals for a single functional," *Biometrika*, **75**, 237–249.

[9] Owen, A.B. (1998) "Empirical likelihood ratio confidence intervals for a single functional," to appear in *The Encyclopedia of Statistics*.

[10] Simon, J.L. (1990). *Resampling Stats*. Arlington, Va.: Resampling Stats, Inc.

[11] Thompson, J.R. and Koronacki, J. (1992). *Statistical Process Control for Quality Improvement*. New York: Chapman & Hall.

[12] Tukey, J.W. (1958). "Bias and confidence in not quite large samples," *Ann. Math. Statist.*, **29**, 614.

[13] Tukey, J.W. (1977). *Exploratory Data Analysis*. Reading, Mass.: Addison-Wesley.

[14] Wilkinson, L. (1998). *SYSTAT 8.0*. Chicago: SPSS, Inc.

Chapter 11

Optimization and Estimation in a Noisy World

11.1 Introduction

In 1949, Abraham Wald [19] attempted to clean up the work of Fisher by proving that the maximum likelihood estimator $\hat{\theta}_n$ of the parameter characterizing a probability density function converged to the true value of the parameter θ_0. He was indeed able to show under very general conditions that if $\hat{\theta}_n$ (globally) maximized the likelihood, it did converge almost surely to θ_0. From a practical standpoint, there is less to the Wald result than one might have hoped. The problem is that, in most cases, we do not have good algorithms for global optimization.

Let us suppose that we seek a minimum to a function $f(x)$. In the minds of many, we should use some variant of Newton's method to find the minimum. Now, Newton's Method does not seek to find the minimum of a function, but rather the (hopefully unique) point where the first derivative of the function is equal to zero. We recall, then, that the simplest of the Newton formulations is an iterative procedure where

$$x_{k+1} = x_k - \frac{f'(x_k)}{f''(x_k)}. \tag{11.1}$$

Returning to Wald's result, let us suppose that we consider data from a Cauchy density

$$f(x) = \frac{1}{\pi[1 + (x - \theta_0)^2]}. \tag{11.2}$$

Then we can pose the maximum likelihood estimation by minimizing the negative of the log likelihood:

$$L(\theta|x_1, x_2, \ldots, x_n) = n\log(\pi) + \sum_{i=1}^{n} \log[1 + (x_i - \theta)^2]. \qquad (11.3)$$

If we use Newton's method here, if we have a starting guess to the left of the smallest point in the data, we will tend to declare that the smallest of the data is the maximum likelihood estimator for θ_0, for that point is a local minimum of the negative of the log likelihood. If we start to the right of the largest data value, we will declare the largest data point to be a maximum likelihood estimator. Now as the data set becomes large, if we do not start on the fringes, we have much better fortune. We do not have multiple "bumps" of the log likelihood near the middle of a large data set. As the data set becomes larger and larger, a starting point that gave a misleading bump at a data far away from θ_0 will, in fact, lead us to an acceptable estimator for θ_0. In one sense, this is true broadly for the problem of estimation by maximum likelihood or minimum χ^2, for example, by Newton's method will become less and less as the sample size increases. We show this stabilization with increasing sample size from a Cauchy distribution with $\theta = 0$ in Figure 11.1. [1] We see, for example, that for the data set explored, for a sample size of 10, we have no false local modes if we start with a positive θ value less than approximately 13, but if we start with a greater value, we might, using Newton's method, wind up with a false maximum (i.e., one that is not equal to the global maximum). Picking any interval of starting values for θ and any values ϵ and δ, however, there will be a sample size such that the probability will be less than ϵ that Newton's method will converge to a value more that δ removed from the true global maximum, namely, θ_0.

This is a phenomenon occurring much more generally than in the case of Cauchy data. In the case of the use of SIMEST in Chapter 5 in the estimation of parameters in a cancer model, for example, the use of the algorithm with sample sizes of 150 demonstrated problems with local maxima, unless one started very near the global maximum. As the number of patients increased past 700, the problem of local maxima of the likelihood (minima of the χ^2) essentially disappeared. That was due to the fact that a larger sample, for maximum likelihood estimation, brings a starting value, unacceptable for smaller samples, into the domain of attraction of the global maximum, and the bumps which existed for the smaller samples, tend to "tail off" (i.e., appear remotely from reasonable starting values). As the sample sizes become large, problems of finding the global maximum of the sample likelihood tend to become less. So, as a practical matter, Wald's result is actually useful if the sample size be sufficiently large.

---

[1] This figure was created by Otto Schwalb.

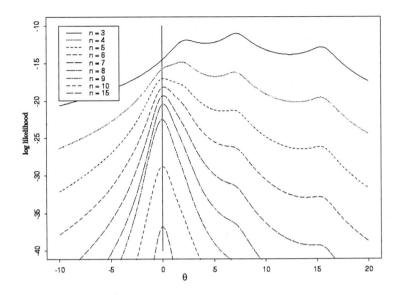

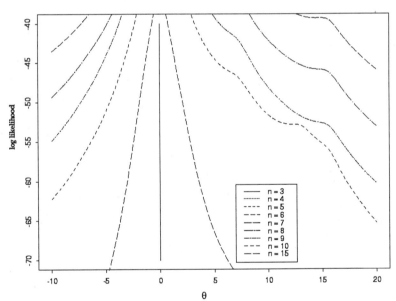

Figure 11.1. Cauchy Log Likelihoods for Various n.

Practical Version of Wald's Result. Let $\{x_1, x_2, \ldots, x_n\}$ be a random sample with a density function $f(x|\theta_0)$ with positive mass throughout its support $a < x < b$. a and/or b may or may not be ∞. For any fixed (i.e, not changing with n) starting point between a and b, as n goes to ∞, if f is well behaved, a Newton's algorithm maximizer of the log likelihood function will converge almost surely to θ_0.

In other words, for the statistician, the natural piling up of data points around regions of high density will cause a practical convergence of the naive maximum likelihood estimator to the truth. Of course, the number of points required to make this conjecture useful may be enormous, particularly if we use it for a multivariate random variable and multivariate characterizing parameter. Statisticians have an advantage over others who deal in optimization, for, generally speaking, the function to be maximized in most problems is not a density function, so that the possibility of never converging to a global maximum is a real one. This, of course, suggests the attractive possibility of trying to reformulate an objective function as a probability density function when feasible to do so. Such a concept is related to that of importance sampling, a topic covered in Section 2.5.

We go further with a conjecture on simulation-based estimation (SIMEST discussed in Chapter 5). Since the time of Poisson [11], it has been taken as natural to model time-based processes in the forward direction. For example, "with a probability proportional to its size, a tumor will, in any time interval, produce a metastasis." Easy to state, easy to simulate—not so easy to find the likelihood, particularly when the metastatic process is superimposed upon other simultaneously occurring processes. For example, the tumor is also growing in proportion to its size; the tumor may be discovered and removed with a probability proportional to its size; and so on. But the simulations superimpose quite readily. Hence, given fast computing, we are tempted to assume the unknown parameters characterizing the pooled processes, generate relevant events by simulation, and then use the difference between the simulated process and the actual, say, discovery of tumors, as a measure of the quality of assumed parameters.

> **SIMEST Conjecture** . Let $\{x_1, x_2, \ldots, x_n\}$ be a random sample with a (possibly not known in closed form) density function $f(x|\theta_0)$. Suppose we can generate, for a given θ, N pseudovalues of $\{y_j\}$ from the density function. Create bins in an intuitive way in the data space, for example putting n/k data points and N_i data points into each of k bins. If the maximum likelihood estimator for θ_0, in the case where we know the closed form of the log likelihood converges to θ_0, then so does an estimator $\hat{\theta}_n$, which maximizes the histogram log likelihood function
>
> $$L_H(\theta|x_1, x_2, \ldots, x_n) = \sum_{i=1}^{k} N_i \log(n/k)$$

based on N pseudodata, as n goes to ∞ and N goes to ∞ if we let k go to ∞ in such a way that $\lim_{n \to \infty} k/n = 0$.

The conjecture concerning simulation-based estimation is rather powerful stuff, since it raises the possibility of parameter estimation in incredibly complex modeling situations. Naturally, for the situation where X is vector valued, some care must be taken in finding appropriate binning strategies (see Section 5.4.2). A somewhat differently styled version of the SIMEST conjecture has been proved by Schwalb [13].

As a matter of fact, the statistician is generally confronted with finding the maximum of an objective function which is contaminated by noise. In the case of simulation-based parameter estimation, the noise is introduced by the modeler himself. The use of Newton-like procedures will generally be inappropriate, since derivatives and their surrogates will be even more unstable than pointwise function evaluation. We shall discuss two ways of dealing with this problem. Interestingly, both the Nelder–Mead algorithm and the Box–Hunter algorithm were built, not by numerical analysts, but by statisticians, working in the context of industrial product optimization. The algorithm of Nelder and Mead essentially gives up on equivalents of "setting the first derivative equal to zero." Rather, it follows an *ad hoc* and frequently very effective zig-zag path to the maximum using pointwise function evaluations without any derivative-like evaluation. The essential idea is to approach the maximum indirectly and, therefore, hopefully, with some robustness. The mighty quadratic leaps toward the maximum promised by Newton's method are not available to the N-M user. On the other hand, neither are the real-world leaps to nowhere-in-particular that frequently characterize Newton's method.

The algorithm of Box and Hunter, on the other hand, takes noisy pointwise function evaluations over a relatively small hyperspherical region and uses them to estimate the parameters of a locally approximating second degree polynomial. Essentially, with Box–Hunter we do take the derivative of the fitting polynomial and proceed to the maximum by setting it equal to zero. But remembering that the fitting validity of the polynomial is generally credible only in a rather small region, we cannot take the giant leaps to glory (or perdition) associated with Newton's method.

11.2   The Nelder–Mead Algorithm

The problem of parameter estimation is only one of many. For most situations, we will not have samples large enough to enable Newton's method to do us much good. Newton method's is generally not very effective for most optimization problems, particularly those associated with data analysis.

A more robust algorithm, one pointing more clearly to the direct search for the minimum (or maximum) of a function, is needed. (And there is still the problem of trying to find the global minimum, not simple some local minimum. We deal with this problem later). Over 30 years ago, two statisticians, Nelder and Mead [10], designed an algorithm which searches directly for a minimum rather than for zeros of the derivative of the function. It does not assume knowledge of derivatives, and it generally works rather well when there is some noise in the pointwise evaluation of the function itself. It is not fast, particularly in higher dimensions. This is due, in part, to the fact that the Nelder–Mead algorithm employs a kind of envelopment procedure rather than one which, as does Newton's method, tries to move directly to the minimum. The Nelder–Mead algorithm is rather intuitive, and, once learned, is easy to construct. We give the algorithm below with accompanying graphs (Figures 11.2 and 11.3) showing the strategy of moving toward the minimum. Our task is to find the minimum of the function $f(x)$. Here, we consider a two-dimensional x.

Nelder–Mead Algorithm
Expansion

- $P = C + \gamma_R(C - W)$(where typically $\gamma_R = \gamma_E = 1$)
- If $f(P) < f(B)$, then
- $PP = C + \gamma_E(C - W)$ [a]
- If $f(PP) < f(P)$, then
- Replace W with PP as new vertex [c]
- Else
- Accept P as new vertex [b]
- End If

Else
If $f(P) < f(2W)$, then

- Accept P as new vertex [b]
- Else

Contraction

If $f(W) < f(P)$, then

- $PP = C + \gamma_C(W - B)$ (typically, $\gamma_C = 1/2$) [a*]
- If $F(PP) < F(W)$, Then replace W with PP as new vertex [b*]
- Else replace W with $(W+B)/2$ and $2W$ with $(2W+B)/2$ (total contraction) [c*]
- End If

Else

Contraction

If $f(2W) < f(P)$, then

- $PP = C + \gamma_C(P - B)$ [aa]
- If $f(PP) < f(P)$, then Replace W with PP as new vertex [bb]

- Else replace W with $(W+B)/2$ and $2W$ with $(2W+B)/2$ (total contraction) [cc]

Else

- Replace W with P
- End If

End If

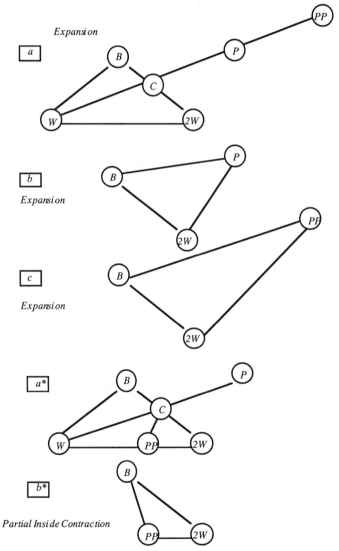

Figure 11.2. Nelder–Mead Polytope Expansions.

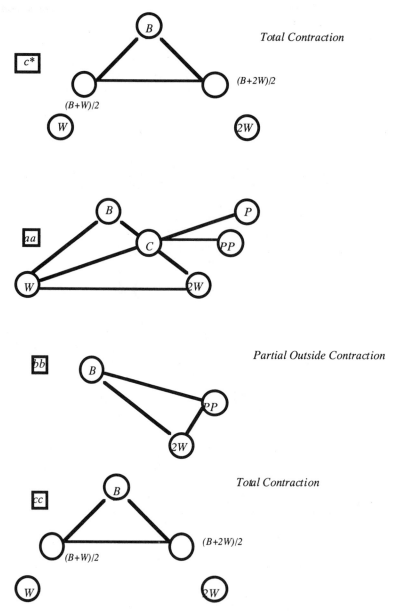

Figure 11.3. Nelder–Mead Polytope Contractions.

11.3   The Box–Hunter Algorithm

Both the Nelder–Mead [10] and Box–Hunter [5] algorithms were designed
with an eye for use in the design of industrial experiments. In fact, they

have been used both as experimental design techniques and as computer optimization routines. However, the Box–Hunter designs clearly are the more used in industry, the Nelder–Mead approach the more used in computer optimization. The reasons are not hard to understand. When it comes to industrial experiments, where great costs are incurred, the rather free-wheeling nature of Nelder–Mead appears profligate. On the computer, where an "experiment" generally simply involves a function evaluation, the Box–Hunter approach appears overly structured, with a "batch" rather than a continuous-flow flavor. However, the natural parallelization possibilities for Box–Hunter should cause us to rethink whether it might be the basis for a new theory of computer optimization and estimation.

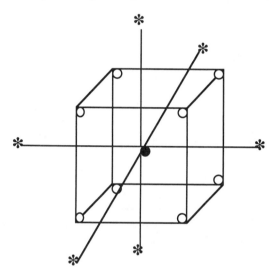

Figure 11.4. Box–Hunter Three-Dimensional Design.

Essentially, the Box–Hunter rotatable design [17] approach centers at the current best guess for the optimum. Points are then placed at the degenerate sphere at that center, the coordinates rescaled so that a movement of one unit in each of the variables produces approximately the same change in the objective function. Then a design is created with points on a hypersphere close to the origin and then on another hypersphere farther out. The experiment is carried out and the coefficients of an approximating quadratic are estimated. Then we move to the new apparent optimum and repeat the process.

In the ensuing discussion, we follow the argument of Lawera and Thompson [9]. The variation of Box–Hunter was created, in large measure, to deal with the application of SIMEST (see Chapter 5) to parameter estimation in stochastic processes. In Figure 11.4, we show a three-dimensional Box–Hunter design. In standardized scaling, the points on the inner hypersphere are corners of the hypercube of length two on a side. The second sphere has "star points" at $2^{p/4}$, where p is the dimensionality of the independent variables over which optimization is taking place. For three dimensions, we have points at $(0,0,0)$, $(1,1,1)$, $(1,1,-1)$, $(1,-1,1)$, $(1,1,-1)$, $(-1,-1,1)$, $(-1,-1,-1)$, $(-1,1,1)$, $(-1,1,-1)$, $(2^{.75},0,0)$, $(-2^{.75},0,0)$, $(0,2^{.75},0)$, $(0,-2^{.75},0)$, $(0,0,2^{.75})$, and $(0,0,-2^{.75})$.

For dimensionality p we start with an orthogonal factorial design having 2^p points at the vertices of the (hyper)cube $(\pm1,\pm1,\ldots,\pm1)$. Then we add $2p$ star points at $(\pm\alpha,0,\ldots,0)$ $(0,\pm\alpha,0,\ldots,0)$, $\ldots$, $(0,0,\ldots,0,\pm\alpha)$. Then we generally add two points (for, say, $p \leq 5$, more for larger dimensionality) at the origin. A sufficient condition for rotatability of the design, i.e., that, as above, $\text{Var}(\hat{y})$ is a function only of

$$\rho^2 = X_1^2 + X_2^2 + \ldots + X_p^2, \tag{11.4}$$

can be shown to be [4] that

$$\alpha = (2^p)^{.25}. \tag{11.5}$$

In Table 11.1 we show rotatable designs for dimensions 2,3,4, 5, 6 and 7.

Table 11.1. Some Box–Hunter Rotatable Designs.				
Dimension	Num. Cube Points	Num. Center Points	Num. Star Points	α
2	4	2	4	$2^{.5}$
3	8	2	6	$2^{.75}$
4	16	4	8	2
5	32	4	10	$2^{1.25}$
6	64	6	12	$2^{1.5}$
7	128	8	14	$2^{1.75}$

When the response variable has been evaluated at the design points, we then use least squares to fit a quadratic polynomial to the results.

$$J_1(\Theta) = \beta_0 + \sum_{i=1}^{p} \beta_i\Theta_i + \sum_{i=1}^{p}\sum_{j=1}^{p} \beta_{ij}\Theta_i\Theta_j. \tag{11.6}$$

We then transform the polynomial to canonical form $\mathbf{A}$:

$$J_2(\Theta) = \beta_0 + \sum_{i=1}^{p} \beta_i\Theta_i + \sum_{i=1}^{p} \beta_{ii}\Theta_i^2. \tag{11.7}$$

Let us now flowchart the Lawera–Thompson version of the Box–Hunter algorithm. First, we define some notation:

Θ_0 coordinates of current minimum

D $n \times 2^n + 2n + n_0$ Box–Hunter design matrix

R $n \times n$ diagonal matrix used to transform into "absolute" coordinate system

T $n \times n$ matrix which rotates the axes of the design to coincide with the "absolute" axes

(Note that the design points as given by matrix D have "absolute" coordinates given by $T \times R \times D + \Theta_0$.)

$S(.)$ the objective function

EC prespecified by the user upper limit on noise level

$CONV$ user-specified constant in the convergence criterion at level 2

X_R $(2^n + 2n + n_0) \times [1 + n + n(n+1)/2]$ matrix of regression points

X can be written as $[\mathbf{1}, (R \times D)^T, rd_{11}, \ldots, rd_{nn}]$, where $\mathbf{1}$ is a column vector obtained by elementwise multiplication of the ith and the jth columns of $(R \times D)^T$.

Lawera–Thompson Algorithm

Level 1
Input (initial guess): Θ_0, R_0, T_0

     1. Perform the level 2 optimization starting from the initial guess. Output: Θ_1, R_1, T_1.

     2. Perform the level 2 optimization 10 times, starting each time from the results obtained in (1).

     3. Find $S_{min}(\Theta_{min})$: the best of results obtained in (2).

Output: $\Theta_{min}, S_{min}(\Theta_{min})$

Level 2
Input: Θ_0, R_0, T_0, EC_0

     1. $EC \leftarrow EC_0$

     2. Perform the level 3 optimization using the input values. Output: $\Theta_1, R_1, T_1, S(\Theta_1)$

     3. Calculate the distance between Θ_0 and Θ_1, i.e., $\Delta\Theta = \sqrt{||\Theta_1 - \Theta_0||^2}$.

     4. Calculate the gain from (2): $\Delta S = S(\Theta_0) - S(\Theta_1)$.

     5. If $\Delta\Theta > 0$ and $\Delta S > 1.5 \times \sqrt{EC}$, then $\Theta_0 \leftarrow \Theta_1, R_0 \leftarrow R_1, T_0 \leftarrow T_1$ and go to (1).

     6. Else if $EC > CONV$, then $EC \leftarrow EC/4$, and go to (2).

     7. Else exit to level 1.

Output: $\Theta_1, R_1, T_1, S(\Theta_1)$.

Level 3
Input: $\Theta_0, R_0, T_0, EC_0, S(\Theta_0)$

1. Perform level 4. Output: R_{min}.

2. Set $Y \leftarrow NULL$, $X \leftarrow NULL$, $i \leftarrow 0$, $\Theta_1 \leftarrow \Theta_0$.

3. Set $R_{Cur} \leftarrow R_{min}$.

4. Increment i by one.

5. Evaluate $S(\Theta)^* = S(T_0 \times R_{Cur} \times D + \Theta_1)$.

6. Calculate $X_{R_{Cur}}$.

7. Set
$$X \leftarrow \left(\begin{array}{c} X \\ X_{R_{Cur}} \end{array} \right).$$

8. Set
$$Y \leftarrow \left(\begin{array}{c} Y \\ S(\Theta)^* \end{array} \right).$$

9. Regress Y on X. Obtain: vector of regression coefficients $\{\hat{\beta}\}$ and the r^2 statistic.

10. Perform the level 5 optimization. Output: Θ^*, R^*, T^*, $S(\Theta)^*$.

11. Calculate the gain from (5): $\Delta S = S(\Theta_1) - S(\Theta)^*$.

12. If $r^2 > 0.9$, $\Delta S > 1.5 \times \sqrt{EC}$, $i < 20$, then
$\Theta_1 \leftarrow \Theta^*$, $R_1 \leftarrow R^*$, $T_1 \leftarrow T^*$
$R_{Cur} \leftarrow 2 \times R_{Cur}$, and go to (4).

13. Else exit to level 2.

Output: Θ_1, R_1, T_1, $S(\Theta_1)$

Level 4
Input: Θ_0, R_0, T_0, EC_0, $S(\Theta_0)$

1. Evaluate $S(\Theta_0)^* = S(T_0 \times R_0 \times D + \Theta_0)$.

2. Calculate X_{R_0}.

3. Regress $S(\Theta_0)^*$ on X_{R_0}. Obtain the r^2 statistic and the error sum of squares (ESS).

4. If $r^2 < 0.9$ and $[ESS < 2 \times EC_0$, or $\mathrm{Max}(S(\Theta_0)) - \mathrm{Min}(S(\Theta_0)) < 1.5 \times \sqrt{EC}]$,

    then

    (a) Set $R_0 \leftarrow 2 \times R_0$.

(b) Repeat (1)–(4) until $r^2 > 0.9$, or $[\,ESS > 2\times EC_0$, and $\mathrm{Max}(S(\Theta_0))- \mathrm{Min}(S(\Theta_0)) < 1.5 \times \sqrt{EC}\,)]$.

(c) Exit to level 3.

5. Else

(a) Set $R_0 \leftarrow 0.5 \times R_0$.

(b) Repeat (1)-(4) until $r^2 < 0.9$, and $ESS < 2 \times EC_0$, or $\mathrm{Max}(S(\Theta_0)) - \mathrm{Min}(S(\Theta_0)) < 1.5 \times \sqrt{EC}$.

Set $R_0 \leftarrow 2 \times R_0$.

(c) Exit to level 3.

Output: R_0

Level 5[2]
Input: $\hat{\beta}$, quadratic fit to the objective function

1. Calculate vector b and matrix B such that the quadratic fit has the form
$$\hat{y} = b_0 + X^T \times b + X^T \times B \times X.$$

2. Find matrices M and Λ such that $M^T \times B = \Lambda$.

3. Calculate the minimum $\Theta \leftarrow -1/2B^{-1} \times b$.

4. If $\sqrt{||\Theta||^2} > 1$, then

(a) Set $\mathrm{Min}(|\Theta_1|,\ldots,|\Theta_n|) \leftarrow 0$.

(b) Repeat (a) until $\sqrt{||\Theta||^2} \leq 1$.

5. Calculate the rescaling matrix $R_0 \leftarrow Diag(|\lambda_i|^{-1/2})$.

6. Set $T_0 \leftarrow M$.

Output: Θ_0, R_0, T_0

Evaluation
Input: Θ, R, T, EC

1. Set $i \leftarrow 0$

2. Evaluate S 10 times at $\Theta_0 = T \times R \times D + \Theta$.

3. Increment i by 7.

4. Calculate the sample mean $\overline{S}$ and the sample variance V of all i evaluations.

---

[2]Level 5 is based on [4].

5. If $V/EC > C_{i-1}$, where C_{i-1} is the 95th percentile of the χ^2_{i-1} distribution, then go to (2).

6. Else exit.

Output $\overline{S}$

Using the Lawera–Thompson variant of the Box–Hunter algorithm on generated tumor data (150 patients) with $\alpha = .31$, $\lambda = .003$, $a = 1.7 \times 10^{-10}$, $b = 2.3a \times 10^{-9}$, using a SIMEST sample size of 1500, with starting value $(.5, .005, 4 \times 10^{-10}, 10^{-9})$, and using 10 bins, we converged to $(.31, .0032, 2 \times 10^{-10}, 2.3 \times 10^{-9})$. Computations were carried out on a Levco desktop parallel processor with 16 CPUs. Subsequent availability of very fast and inexpensive serial machines has caused us, temporarily, to suspend the parallel investigation. It is clear, however, that the Box–Hunter paradigm, requiring minimal handshaking between CPUs, is a natural candidate for parallelization.

11.4   Simulated Annealing

The algorithm of Nelder–Mead and that of Box–Hunter will generally not stall at a local minimum with the same degree of risk as a Newton's method based approach. Nevertheless, it sometimes happens that stalling local minima do occur with Nelder–Mead. How to get around this problem?

Naturally, there is no easy answer. Practically speaking, there is no general way to make sure that a minimum is global without doing a search over the entire feasible region. We might well converge, using Nelder–Mead to point **A** in Figure 11.5. We need some way to make sure that we really have arrived at the global minimum.

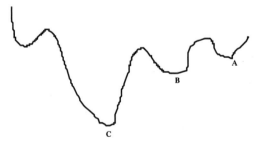

Figure 11.5. The Problem with Local Minima.

In the example in Figure 11.5, we might argue that we need something to kick us away from **A**. Then, we can see if, say using Nelder–Mead, we move to a new candidate for the global minimum, say **B**, or fall back to **A**. From the picture, it is clear that it is a critical matter just how far we move away from **A** which determines whether we will progress on to **B** and thence to

C, or whether we will fall back into **A**. Mysterious analogies to Boltzmann energy levels are probably not very helpful, but we shall mention the idea, since users of *simulated annealing* generally do.

In cooling a molten metal too quickly, one may not reach a level of minimum energy (and hence apparently of crystalline stability). When this happens, it sometimes happens that a decision is made to reheat the metal (though not to the molten state, necessarily) and then cool it down again, slowly, hoping to move to another lower-energy state. The probability of moving from energy state E_1 to another energy state E_2, when $E_2 - E_1 = \Delta E > 0$, is given by $\exp(-\Delta E/(kT))$ where k is the *Boltzmann constant*. The analogy is relatively meaningless, and generally the $1/(kT)$ is simply replaced by a finagle factor. The finagle factor will determine how far we kick away from the apparent minimum. As time progresses, we may well decide to "stop the kicking."

Let us suppose that we have used Nelder–Mead to get to a minimum point x_0. Using this as our starting point, we can use an algorithm suggested by Bohachevsky, Johnson, and Stein [1, 12].

BJS "General" Simulated Annealing Algorithm

1. Set $f_0 = f(x_0)$. If $|f_0 - f_m| < \epsilon$, stop.

2. Generate n independent standard normal variates $Y_1, Y_2, \ldots, Y_n$ Let $U_i = Y_i/(Y_1^2 + Y_2^2 + \ldots + Y_n^2)^{1/2}$ for $i = 1, 2, \ldots, n$.

3. Set $x^* = x_0 + (\Delta r)U$.

4. If x^* is not in the feasible set, return to step 2, otherwise, set $f_1 = f(x^*)$ and $\Delta f = f_1 - f_0$.

5. If $f_1 \leq f_0$, set $x_0 = x^*$ and $f_0 = f_1$. If $|f_0 - f_m| < \epsilon$, stop. Otherwise, go to step 2.

6. If $f_1 > f_0$, set $p = \exp(-\beta f_0^g \Delta f)$.

7. Generate a uniform $\mathcal{U}(0,1)$ random variate V. If $V \geq p$, go to step 2. If $V < p$, set $x_0 = x^*$, $f_0 = f_1$ and go to step 2.

It is clear that the above algorithm contains a fair amount of things which are a bit arbitrary. These include

- Step size Δr

- g

- β

- f_m assumed value of the global minimum

Before the advent of simulated annealing, investigators tried to seek pathways to the global optimum by starting at a random selection of starting points. Such a *multistart* approach is still a good idea. Taking a set of local minima obtained from different starting points, one might try a number of strategies of starting from each of the local minima and conducting a random search in hyperspheres around each to see if better minima might be obtained, and so on. Any simulated annealing approach will be, effectively, a random search on a set much smaller than the entire feasible region of the parameter space. We should despair, in general, of coming up with a foolproof method for finding a global optimum that will work with any and all continuous functions. Much of the supposed success of simulated annealing, as opposed to the kind of multistart algorithm, is probably due to the very fast computers that simulated annealers tended to have available.

11.5   Exploration and Estimation in High Dimensions

The power of the modern digital computer enables us realistically to carry out analysis for data of higher dimensionality. Since the important introduction of exploratory data analysis in the 1970s, a great deal of effort has been expended in creating computer algorithms for visual analysis of data. One major advantage of EDA compared to classical procedures is a diminished dependency on assumptions of normality. However, for the higher-dimensional situation, visualization has serious deficiencies, since it tends to involve projection into two or three dimensions.

What are typical structures for data in high dimensions? This is a question whose answer is only very imperfectly understood at the present time. Some possible candidates are:

1. Gaussian-like structure in all dimensions.

2. High signal-to-noise ratio in only in one, two, or three dimensions, with only noise appearing in the others. Significant departures from Gaussianity.

3. System of solar systems. That is, clusters of structure about modes of high density, with mostly empty space away from the local modes.

4. High signal-to-noise ratio along curved manifolds. Again the astronomical analogy is tempting, one appearance being similar to that of
spiral nebulae.

For structure 1, classical analytical tools are likely to prove sufficient. For
structure 2, EDA techniques, including nonparametric function estimation
and other nonparametric procedures will generally suffice. Since human
beings manage to cope, more or less, using procedures which are no more
than three- or four-dimensional, it might be tempting to assume that stuc-
ture 2 is somehow a natural universal rule. Such an assumption would
be incredibly anthropomorphic, and we do not choose, at this juncture, to
make it. For structure 3, the technique investigated by Thompson and his
students [2, 6−8] is the finding of modes, utilizing these as base camps for
further investigation. For structure 4, very little successful work has been
done. Yet the presence of such phenomena as diverse in size as spiral nebulae and DNA shows that such structures are naturally occurring. One way
in which the astronomical analogy is deceptively simple is that astronomical problems are generally concerned with relatively low dimensionality.
By the time we get past four dimensions, we really are in *terra incog-
nita* insofar as the statistical literature is concerned. One hears a great
deal about the "curse of dimensionality." The difficulty of dealing with
higher-dimensional non-Gaussian data is currently a reality. However, for
higher-dimensional Gaussian data, knowledge of data in additional dimensions provides additional information. So may it also be for non-Gaussian
data, if we understood the underlying structure.

Here, we are concerned mainly with structure 3. Mode finding is based
on the mean update algorithm (MUA) [2 , 6, 7, 18]:

mean update algorithm
Let $\hat{\mu}_1$ be the initial guess
Let m be a fixed parameter;
$i = 1$;
Repeat until $\hat{\mu_{i+1}} = \hat{\mu}_i$;
Begin
Find the sample points $\{X_1, X_2, \ldots, X_m\}$ which are closest to μ_i;
Let $\hat{\mu_{i+1}} = \frac{1}{m} \sum_{j=1}^{m} X_j$;
$i = i + 1$;
end.

Let us consider a sample from a bivariate distribution centered at $(0,0)$.
The human eye easily picks the $(0,0)$ point as a promising candidate for the
"location" of the distribution. Such a Gestaltic visualization analysis is not
as usable in higher dimensions. We will be advocating such an automated
technique as the mean update algorithm. Let us examine Figure 11.6.
Suppose that we have only one dimension of data. Starting at the projection
of **0** on the x-axis, let us find the two nearest neighbors on the x-axis.

Taking the average of these, brings us to the **1** on the x-axis. And there the algorithm stalls, at quite a distance from the origin.

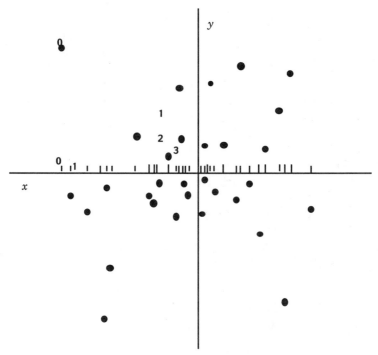

Figure 11. 6. Mean Update Estimation of Mode.

On the other hand, if we use the full two dimensional data, we note that the algorithm does not stall until point **3**, a good deal closer to the origin. So increased dimensionality need not be a curse. Here, we note it to be a blessing.

Let us take this observation further. Suppose we are seeking the location of the minor mode in a data set which (unbeknownst to us) turns out to be

$$f(x) = .3\mathcal{N}(x; .051, \mathbf{I}) + .7\mathcal{N}(x; 2.4471, \mathbf{I}). \qquad (11.8)$$

If we have a sample of size 100 from this density and use the mean update algorithm, we can measure the effectiveness of the MUA with increasing dimensionality using the criterion function

$$\mathrm{MSE}(\hat{\mu}) = \frac{1}{p} \sum_{j=1}^{p} (\hat{\mu}_j - \mu)^2. \qquad (11.9)$$

Below we consider numerical averaging over 25 simulations, each of size 100.

Table 11.2. Mean Square Errors		
p	m	MSE
1	20	.6371
3	20	.2856
5	20	.0735
10	20	.0612
15	20	.0520

We note in Table 11.2, how, as the dimensionality increases, essentially all of the 20 nearest neighbors come from the minor mode, approaching the idealized MSE of .05 as p goes to ∞. Subsequent work [6] has shown that for multiple modes in dimensions five and over, the MUA appears to find, automatically, which points to associate with each mode, so that even for mixtures of rather taily distributions such as $T(3)$, we come close to the idealized MSE for the location of each mode, namely, $1/(np)$ where n is the total sample size and p is the proportion of the data coming from the mode. So far from being a curse, an increasing dimensionality can be an enormous blessing. We really have no very good insights yet as to the what happens in, say, 8-space. This examination of higher-dimensional data is likely to be one of the big deals in statistical analysis for the next 50 years. The examination of data in higher dimensions is made possible by the modern computer. If we force ourselves, as is currently fashionable, to deal with higher-dimensional data by visualization techniques (and hence projections into 3-space) we pay an enormous price and, quite possibly, miss out on the benefits of high dimensional examination of data.

The analogy we shall employ is that of moving through space until we find a "center" of locally high density. We continue the process until we have found the local modes for the data set. These can be used as centers for local density estimation, possibly nonparametric, possibly parametric (e.g., locally Gaussian). It turns out, as we shall see, that finding local modes in high dimensions can be achieved effectively with sample sizes orders of magnitude below those generally considered necessary for density estimation in high dimensions [14−16]. Moreover, as a practical matter, once we have found the modes in a data set, we will have frequently gleaned the most important information in the data, rather like the mean in a one-dimensional data set.

Let us suppose that we have, using each data point from the data set of size n as a starting point, found mm apparent local modes. As a second step, let us develop an algorithm for consolidating the apparent local modes to something more representative of the underlying distribution. There are many ways to carry out the aggregation part of the algorithm. This is only one of the possibilities.

Take two of the local modes, say M_1 and M_2. Examine the volume $V_{1,m}$ required to get, say, m nearest neighbors of M_1 and $V_{2,m}$ required to get, say, m nearest neighbors of M_2. Standing at the midpoint between M_1 and M_2, say, $M_{1,2}$, draw a sphere of volume $V_{12,m} = V_{1,m} + V_{2,m}$. Suppose that the number of distinct points in the pooled clouds is m_{12}. Suppose that the hypersphere centered at $M_{1,2}$ has a density as high as that in the other two clouds. Let the number of points falling inside the hypersphere centered at $M_{1,2}$ be $n_{1,2}$. Then if the number of data points falling inside that hypersphere is greater than the 5th percentile of a binomial variate of size m_{12} and with $p = 0.5$, we perform a condensation step by replacing the two modes M_1 and M_2 by $M_{1,2}$ as shown in Figure 11.7.

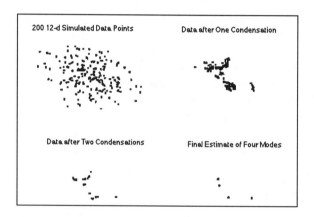

Figure 11.7. Condensation Progression Using MUA.

To examine the progression of the condensation algorithm, we simulate 200 data points from a mixture of four 12-dimensional normal distributions with mixture weights .40, .24, .19, and .17. The four modes are well estimated both in terms of numerosity and location even for such a small data set.

Let us apply Elliott's version [6, 7] of the MUA to the much-studied Fisher−Anderson iris data. This is a database of three varieties of iris with 50 observations from each of the varieties. The algorithm found four (see Table 11.4) rather than the hoped-for three clusters shown in Table 11.3.

Table 11.3. Fisher−Anderson Iris Data.				
Species	Sepal Length	Sepal Width	Petal Length	Petal Width
Setosa	5.006	3.428	1.462	0.246
Versicolor	5.936	2.770	4.260	1.326
Virginica	6.588	2.974	5.552	2.026

Table 11.4. Estimated Modes Based on 150 Observations				
Species	Sepal Length	Sepal Width	Pet. Lgth	Pet. Width
Setosa	4.992	3.411	1.462	.225
Versicolor	5.642	2.696	4.101	1.267
Virginica	6.762	3.067	5.589	2.218
Versi/Virgin.	6.249	2.893	4.837	1.591

At first, we feared that some fundamental flaw had crept into the algorithm, which had always performed quite predictably on simulated data. Later, it seemed plausible, based on the fact *Verginica* and *Versicolor* always spontaneously hybridize and that it is almost impossible not to have this hybrid present, to believe our eyes. The fourth mode had occurred, in fact, almost precisely at the mean of the *Verginica* and *Versicolor* modes. This was a rather surprising result, apparently unnoticed in the some fifty years since the Fisher−Anderson iris data became something of a test bed for measuring the effectiveness of discrimination and clustering algorithms.

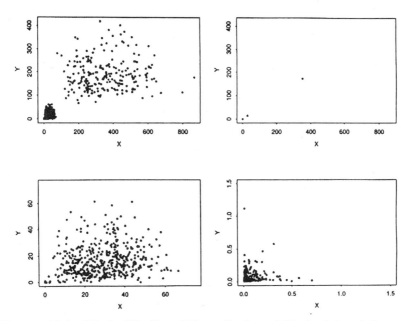

Figure 11.8. Mode Finding When Scales of Underlying Mixtures Are Very Different.

In another application, Elliott and Thompson [6] have examined a four dimensional ballistics data set of size 944 kindly provided by Malcolm Taylor of the Army Research Laboratory. Consider the two-dimensional projections displayed in Figure 11.8. Our algorithm was able to find modes from overlapping subpopulations whose scales differed by nearly 1000. We see in the top left quadrant of Figure 11.8 a two-dimensional projection of the data set. The top right quadrant gives the three estimated modes. In the lower left quadrant, we have zoomed in on the cluster in the lower left of the data set. In the lower right quadrant we have zoomed in to a scale

10^{-3} of that used in the display of the raw data. As we have seen, even in data sets of dimensionality as low as four, there seems to be appearing a big bonus for the extra dimension(s) past three for finding modes.

Mean update algorithms show great promise for exploratory purposes. The problem of nonparametric function estimation is one to which some of us at Rice have given some attention for a number of years. Our foray into the higher dimensions has produced a number of surprises. The notion that increasing dimensionality is a "curse" seems only to be true if we insist on graphical approaches. Our multidimensional mode–finding algorithm dramatically improves with increasing dimensionality.

Problems

11.1. In optimization examples, perhaps the easiest problem is that of finding the minimum of the dot product. Consider finding the minimum of

$$J_1(\Theta) = \Theta_1^2 + \Theta_2^2 + \Theta_3^2 + 2\epsilon$$

where ϵ is $\mathcal{N}(0,1)$. Examine the performance of both the Nelder–Mead and Box–Hunter algorithms.

11.2. A somewhat more difficult minimization case study is that of the Rosenbrock function with additive Gaussian noise

$$J_2(\Theta) = 100(\Theta_1^2 - \Theta_2^2)^2 + (1 - \Theta_1)^2 + 1 + \epsilon,$$

where ϵ is $\mathcal{N}(0,1)$. Examine the performance of both the Nelder–Mead and Box–Hunter algorithms.

11.3. Returning to the problem in Section 5.4.1, generate a set of times of discovery of secondary tumor (time measured in months past discovery and removal of primary) of 400 patients with $a = .17 \times 10^{-9}$, $b = .23 \times 10^{-8}$, $\alpha = .31$, and $\lambda = .0030$. Using SIMEST, see if you can recover the true parameter values from various starting values, using the Box–Hunter algorithm.

11.4. Consider the density function

$$f(x) = .5\mathcal{N}(x; .551, \mathbf{I}) + .3\mathcal{N}(x; 21, \mathbf{I}) + .2\mathcal{N}(x; 21, \mathbf{I})$$

Generate random samples of size 100 for dimensions 2, 3, 4, 5, and 10. Examine the efficacy of the MUA in finding the centers of the three distributions.

References

[1] Bohachevsky, I.G., Johnson, M.E., and Stein, M.L. (1988). "Generalized simulated annealing for function optimization," *Technometrics*, 209–217.

[2] Boswell, S.B. (1983). *Nonparametric Mode Estimation for Higher Dimensional Densities*. Doctoral dissertation. Houston: Rice University.

[3] Box, G.E.P. and Draper, N.R. (1969). *Evolutionary Operation*. New York: John Wiley & Sons.

[4] Box, G.E.P. and Draper, N.R. (1989). *Empirical Model-Building and Response Surfaces*. New York: John Wiley & Sons.

[5] Box, G.E.P. and Hunter, J.S. (1957). "Multifactor experimental designs for exploring response surfaces," *Ann. Statist.*, **28**, 195–241.

[6] Elliott, M.N. and Thompson, J.R. (1993). "The nonparametric estimation of probability densities in ballistics research," *Proceedings of the Twenty-Sixth Conference of the Design of Experiments in Army Research Development and Testing*. Research Triangle Park, N.C.: Army Research Office, 309–326.

[7] Elliott, M.N. (1995). *An Automatic Algorithm For The Estimation of Mode Location and Numerosity in General Multidimensional Data*. Doctoral dissertation. Houston: Rice University.

[8] Fwu, C., Tapia, R.A, and Thompson, J.R. (1981). "The nonparametric estimation of probability densities in ballistics research," *Proceedings of the Twenty-Sixth Conference of the Design of Experiments in Army Research Development and Testing*. Research Triangle Park, N.C.: Army Research Office, 309–326.

[9] Lawera, M. and Thompson, J.R. (1993). "A parallelized, simulation based algorithm for parameter estimation," *Proceedings of the Thirty-Eighth Conference on the Design of Experiments in Army Research Development and Testing*, B. Bodt, ed. Research Triangle Park, N.C.: Army Research Office, 321–341.

[10] Nelder, J.A. and Mead, R. (1965). "A simplex method for exploring response surfaces," *Comput. J.*, 308-313.

[11] Poisson, S.D. (1837). *Recherches sur la probabilité des jugements en matière criminelle et en matière civile, précedées des réglés générales du calcul des probabilités*. Paris.

[12] Press, W.H., Teukolsky, S.A., Vetterling, W.T., and Flannery, B.P. (1992). *Numerical Recipes in FORTRAN*. New York: Cambridge Univer-

sity Press, 443–448.

[13] Schwalb, O. (1999). *Practical and Effective Methods of Simulation Based Parameter Estimation for Multidimensional Data.* Doctoral dissertation. Houston: Rice University.

[14] Scott, D.W. (1992). *Multivariate Density Estimation.* New York: John Wiley & Sons, 199.

[15] Scott, D.W. and Thompson, J.R. (1983). "Probability density estimation in higher dimensions," in *Computer Science and Statistics*, J. Gentle, ed. Amsterdam: North Holland, 173–179.

[16] Silverman, B.W. (1986). *Density Estimation for Statistics and Data Analysis.* London: Chapman & Hall.

[17] Thompson, J.R. and Koronacki, J. (1993). *Statistical Process Control for Quality Improvement.* New York: Chapman & Hall, 238–245.

[18] Thompson, J.R. and Tapia, R.A. (1990). *Nonparametric Function Estimation, Modeling, and Simulation.* Philadelphia: SIAM, 167–168.

[19] Wald, A. (1949). "Note on the consistency of the maximum likelihood estimate," *Ann. Math. Statist.*, 595–601.

Chapter 12

Modeling the AIDS Epidemic: Exploration, Simulation, Conjecture

12.1 Introduction

We started this book discussing the primary function of simulation as a modeling of reality. Data are not valuable *sui generis*, only as they point to the system which generated them. Many millions of dollars have been spent on "analysis" of AIDS cases in the United States, but the analyses have been generally treated without regard to the possibility of impact on public policy. The epidemic has already claimed over 400,000 American lives.

As is the case with many important societal problems, the reasons for the AIDS catastrophe may never be known for certain. Early on in the AIDS epidemic, Thompson [3] developed a model demonstrating that a small subpopulation, with unusually high rates of sexual activity, in the American gay community could serve to drive the endemic across the epidemiological threshold into a full epidemic. Over 15 years after its formulation, the model still holds up.

At the time of the original work on the model, there seemed to be no reason to utilize a stochastic model; a deterministic model appeared to suffice. Extensive subsequent work by West [9] indicates that indeed the essentials of the epidemiological phenomenon of AIDS can be handled with a straightforward system of differential equations. Such a model is a departure from most of the rest of this book. However, as noted in the preface, deterministic model-based simulation is frequently satisfactory, depending on the situation. In the case of epidemics, examination of the mean trace will usually suffice.

In the case of AIDS, assuming that the agent of infection was of ancient origin, it was desirable to ask what sociological changes had taken place to render the epidemic feasible. One aspect of HIV appears to be that virus in some quantity is usually required for transmission. Thus, as a venereal disease, female-to-male transmission is rather difficult. Since the disease, in the United States, appeared first in the homosexual male community, that would be consistent with the low infectivity of the virus. But, that left us with the question as to why the epidemic had not appeared in the early 1970s as opposed to the early 1980s, for the societal prohibitions against homosexual behavior had largely disappeared in the United States around 1970. One reasonable way to proceed was to look at sociological changes which had taken place from, say, 1975, on. This was the time when municipalities in the United States started to treat the gay bathhouses as social clubs rather than as places of illicit sexual activity. In the 1984 paper [3], a look was taken to see what could happen as the result of a subpopulation in the American gay male community which had sexual activity well about the average. Numerical solution of the differential equations model showed that the bathhouses were, indeed, a prime candidate for the cause of the AIDS epidemic in the United States. This effect was not due to the aggregate increase of sexual activity in the gay community, but rather to the effect of the small subpopulation [4–6].

From a public health standpoint, the model indicated that it would be appropriate for centers of high-contact anonymous sex, such as the gay bathhouses, to be closed. Except for short periods of time in random locations, such closings have not been implemented by the U.S. public health authorities. Thompson has drawn comparisons with the ineffectiveness of public health policy in the American polio epidemic in the 1940s [5]. In the case of the polio epidemic, it was fairly clear that closings of public swimming pools and cheap Saturday afternoon matinees were indicated, but public health officials did not wish to take the static for such actions.

More recent work by Thompson and Go [7] indicates that once the percentage of HIV infectives in a local gay population reaches 40%, the benefit of such closings is probably marginal. There is little doubt that such a rate has been reached already in a number of U.S. urban centers, though probably not in all. At any rate, U.S. public health policy in the management of the AIDS epidemic has proved catastrophic. Most of the money spent to date in AIDS research comes from U.S. sources. Nevertheless, we continue to have much the highest AIDS rate per 100,000 in the First World, and recent WHO statistics reveal that the AIDS rate in the United States is over five times that of the rest of the First World. Houston alone has more AIDS cases than in all of Canada (which has more than ten times Houston's population). Recent work by West and Thompson [9–11] shows that the weakened resistance of HIV infectives makes them more susceptible to tuberculosis. Since tuberculosis is an aerosol-borne disease, West [9] investigated the possibility that the number of infections into the non-HIV population could have dramatic public health consequences. Fortunately, his work indicates that this is unlikely, with a worst-case scenario of under 10,000 additional TB cases per year for the entire nation. His model indicates that modest additional funding to tuberculosis treatment centers might bring the marginal increase in TB due to HIV infectives to no more than a few hundred.

Notions that most persons engaging for long times in high-risk behavior will be eliminated by the epidemic seem unfortunately to be true in the aggregate. However, notions that the epidemic simply will end as a result of the removal by the disease of these infectives from the population may not be correct, due to a continuing supply of new susceptibles. It is entirely possible that, absent a cure or a vaccine, AIDS could have devastating effects in the United States until well into the next century.

12.2   Current AIDS Incidences

In the matter of the present AIDS epidemic in the United States, a great deal of money is being spent on AIDS. However, practically nothing in the way of steps for stopping transmission of the disease is being done (beyond education in the use of condoms). Indeed, powerful voices in the Congress speak against any sort of government intervention. On April 13, 1982, Congressman Henry Waxman [2] stated in a meeting of his Subcommittee on Health and the Environment, "I intend to fight any effort by anyone at any level to make public health policy regarding Kaposi's sarcoma or any other disease on the basis of his or her personal prejudices regarding other people's sexual preferences or life styles." (It is interesting to note that Representative Waxman is one of the most strident voices in the fight to stop smoking, considering rigorous measures acceptable to end this threat to human health.) We do not even have a very good idea as to what fraction of the target population in the United States is HIV positive, and anything approaching mandatory testing is regarded by U.S. political leaders as an unacceptable infringement on civil liberties.

In Table 12.1, we show new AIDS figures for six First World Countries.[1]

---

[1]The author wishes to express his gratitude to Ms. Rachel Mackenzie and the other members of the Working Group on Global HIV/AIDS and STD Surveillance which is a joint Working Group between WHO and UNAIDS.

Table 12.1. First World AIDS Figures (New Cases)						
Year	U.S.A.	U.K.	Canada	France	Denmark	Netherlands
1984*	10,599	161	265	371	36	55
1985	11,386	244	371	583	38	67
1986	18,677	470	630	1257	69	136
1987	28,573	675	946	2246	100	242
1988	36,715	900	1146	3047	126	325
1989	43,273	1366	4342	3799	174	391
1990	50,031	1410	5462	4040	197	418
1991	59,284	1538	7164	4643	210	446
1992	77,984	1709	8624	5165	209	509
1993	77,641	1725	9914	5477	239	469
1994	68,959	1671	10391	5945	235	490
1995	61,614	1433	1290	5141	214	486
*These entries are cumulative totals through 1984.						

We note how large are the figures from the United States. Since AIDS is generally a fatal disease, it is useful to compare US AIDS counts with US dead from the wars it has fought in Table 12.2.

Table 12.2. U.S. War Deaths	
Conflict	U.S. Combatant Deaths
Revolutionary War	6,000
War of 1812	4,000
War with Mexico	2,000
War Between the States	550,000
Spanish-American War	3,000
World War I	120,000
World War II	410,000
Korean War	55,000
War in Southeast Asia	60,000
Cumulative U.S. AIDS cases (mid-1997)	612,000
Estimated cumulative US HIV positives	1,000,000

In Figure 12.1, we note the staggering difference in cumulative AIDS cases between the United States and France, Denmark, Netherlands, Canada, and the United Kingdom. The pool of infectives in the USA simply dwarfs those of the other First World countries.

Shall we simply note, without undue interest that AIDS is an order of magnitude higher in the United States when compared to the average of other First World countries? From a nominalist standpoint, such an intellectual shrug of the shoulders may be acceptable. But for the modeler, there is an imperative to try and find out what is going on.

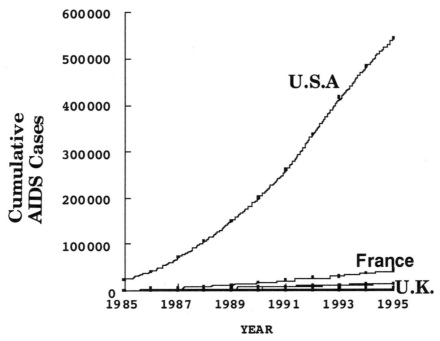

Figure 12.1. Cumulative AIDS Cases 1985-1995.

12.3 Discussion

Let us see what one might expect if other First World countries were lagging the United States. Then one would expect some sort of variation of Figure 12.2. That simply is not what is happening as we see in Figure 12.3. No other First World country is catching up to the United States. Moreover, a downturn in new case rates is observable in all the countries shown. Further insight is given in Figure 12.4 where we divide the annual incidence of AIDS per 100,000 in the United States by that for various other First World countries.

There seems to be a relative constancy, across the years, of the new case ratio of that of the United States when divided by that for each country shown. Thus, for the United Kingdom, it is around 9, for Denmark 6, and so on. It is a matter of note that this relative constancy of new case rates is maintained over the period examined (11 years).

If the USA Simply Leads the Rest of the First World in AIDS

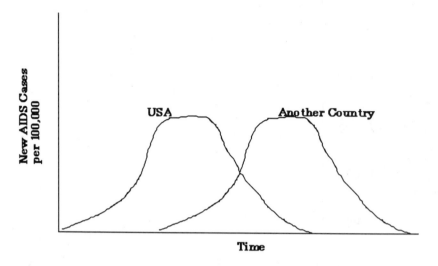

Figure 12.2. A Time-Lagged Scenario.

New Case Rates

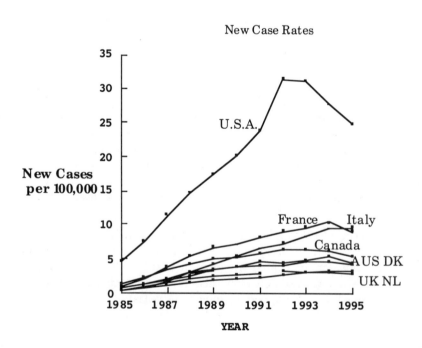

Figure 12.3. New Case Rates per 100,000.

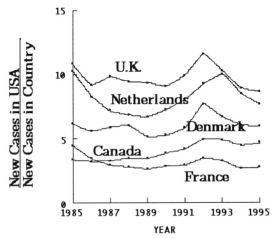

Figure 12.4. Comparative New Case Rates.

We notice, in Figure 12.5, that the cumulative instance per 100,000 of AIDS in the USA divided by that for other First World countries gives essentially the same values as shown for the new case rates in Figure 12.4.

Next, let us consider the piecewise in time exponential model for the number of AIDS cases in, say, Country A.

$$\frac{dy_A}{dt} = k_A y_A. \tag{12.1}$$

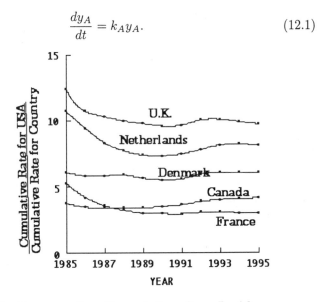

Figure 12.5. Comparative Cumulative Case Incidence.

We show in Figure 12.6 estimates for k rates on a year-by-year basis using

$$k_A(t) \approx \frac{\text{new cases per year}}{\text{cumulative cases}}. \tag{12.2}$$

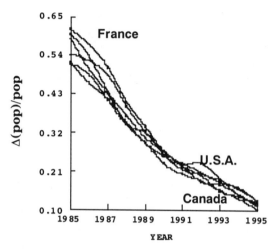

Figure 12.6. k **Values for Canada, Denmark, France, Netherlands, the United Kingdom, and the United States.**

We note the apparent near equality of rates for the countries considered. To show this more clearly, in Figure 12.7, we show the ratio of the annual estimated piecewise national rates divided by the annual estimated rate of the United States.

It is a matter of some interest that k values are essentially the same for each of the countries shown in any given year. How shall we explain a situation where one country has a much higher incidence of new cases, year by year, and the rate of increase for all the countries is the same? For example, by mid-1997, the United Kingdom had a cumulative total of 15,081 cases compared to the United States' 612,078. This ratio is 40.59, whereas the ratio of populations is only 4.33. This gives us a comparative incidence proportion of 9.37. On the other hand, at the same time, Canada had cumulative AIDS total of 15,101. The U.S. population is 9.27 times that of Canada. The comparative incidence proportion for the United States versus Canada in mid-1997 was 4.37. The comparative incidence of the United States vis-à-vis the United Kingdom is over twice that of the United States vis-à-vis Canada. Yet, in all three countries the rate of growth of AIDS cases is nearly the same. This rate changes from year to year, from around .54 in 1985 to roughly .12 in 1995. Yet it is very nearly the same for each country in any given year. One could, therefore, predict the number of new cases in France, in a given year, just about as well knowing the case history of the United States instead of that in France. The correlation of new cases for the United States with that for each of the other countries considered is extremely high, generally around .96. Can we explain this by an appeal to some sort of magical synchronicity? I think not. Particularly since we have the fact that although the growth rates of AIDS for the countries are roughly the same for any given year, the new case relative incidence per 100,000 for the United States is several times that of any of the other countries.

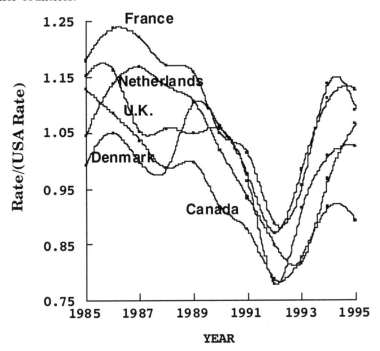

Figure 12.7. Ratios of Piecewise Rate Estimates for Canada, Denmark, France, the Netherlands, and the United Kingdom Relative to That of the United States.

Earlier in this chapter we noted the conjecture I made in the mid-1980s that it was the bathhouses which caused the stand-alone epidemic in the United States. But, as we have seen, the bathhouse phenomenon really does not exist in the rest of the First World. How is it, then, that there are stand-alone AIDS epidemics in each of these countries? I do not believe there are stand-alone AIDS epidemics in these countries.

In Figure 12.8, let us suppose there is a country, say Country Zero, a country where, for whatever reason, the sociology favors a stand-alone AIDS epidemic. From other First World countries there is extensive travel to and from Country Zero. If AIDS, with its very low infectivity rates, break, out in Country Zero, then, naturally, the disease will spread to other countries. But, if the intracountry infectivity rate is sufficiently low, then the maintenance of an apparent epidemic in each of the countries will be dependent on continuing visits to and from Country Zero.

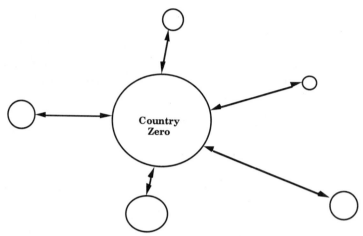

Figure 12.8. Country Zero.

Now let us suppose that the fraction of infectives is rather low in Country j. Thus, we shall assume that the susceptible pool is roughly constant. Let the number of infectives in Country j be x_j. Let the number of infectives in Country Zero be given by z. Suppose the case rate in Country Zero divided by that for Country j is relatively constant, say c_j. Let us suppose that we have the empirical fact that, for both Country Zero and the other countries, we can use the same β_t in

$$\frac{dz}{dt} = \beta_t z \tag{12.3}$$

$$\frac{dx_j}{dt} = \beta_t x_j. \tag{12.4}$$

Let us suppose that, at any given time, the transmission of the disease in a country is proportional both to infectives in the country and to infectives in Country Zero. Let us suppose that the population of infectives in Country j is given by x_j and in Country j is z. Let the population of Country j be given by N_j and that of Country Zero be given by N_Z. Suppose that we have, at any time, the following constancy

$$\frac{z/N_Z}{x_j/N_j} = c_j. \tag{12.5}$$

Then we have

$$
\begin{aligned}
\frac{dx_j}{dt} &= \alpha_{j,t} x_j + \eta_{j,t} z \\
&= \left(\alpha_{j,t} + \frac{N_Z}{N_j} c_j \eta_{j,t} \right) x_j \\
&= \beta_t x_j. \tag{12.6}
\end{aligned}
$$

where $\alpha_{j,t}$ and $\eta_{j,t}$ are the transmission rates into Country j from the country infectives and the Country Zero infectives, respectively. Now, we are assuming that the effect of infectives from the countries will have relatively little effect on the increase of infectives in Country Zero. Thus, for a short time span,

$$z(t) \approx z(0)e^{\beta_t t}. \tag{12.7}$$

Thus (12.6) is roughly

$$\frac{dx_j}{dt} = \alpha_{j,t}x_j + \eta_{j,t}z(0)e^{\beta_t t}. \tag{12.8}$$

Now we note that the epidemic in a country can be sustained even if $\alpha_{j,t}$ is negative, provided that the transmission from the Country Zero infectives is sufficiently high. If we wish to look at the comparative effect of Country Zero transmission on country j vis-à-vis country k, we have

$$\eta_{j,t} = \frac{c_k}{c_j}\frac{N_j}{N_k}\eta_{k,t} + \frac{\alpha_{k,t} - \alpha_{j,t}}{c_j}\frac{N_j}{N_k}. \tag{12.9}$$

Let us suppose that for two countries, we have

$$\alpha_{j,t} = \alpha_{k,t}. \tag{12.10}$$

Then we have

$$\eta_{j,t} = \frac{c_k}{c_j}\frac{N_j}{N_k}\eta_{k,t}. \tag{12.11}$$

Looking at this another way, we have

$$\frac{x_j}{x_k} = \frac{\eta_j}{\eta_k}. \tag{12.12}$$

If η_j doubles, then according to the model, the number of infectives in country j doubles. Let us see what the situation would be in Canada if, as a stand-alone, the epidemic is just at the edge of sustainability, that is,

$$\alpha_{\text{Can}} = 0. \tag{12.13}$$

Then, going back to a universal β_t for all the countries and Country Zero America as well, we have

$$\begin{aligned}
\eta_{\text{Can},t} &= \frac{N_{\text{Can}}}{N_{USA}}\frac{1}{c_{\text{Can}}}\beta_t \\
&= \frac{26,832,000}{248,709,873}\frac{1}{4.14}\beta_t \\
&= .026\beta_t.
\end{aligned} \tag{12.14}$$

Thus, according to the model, activity rates from U.S.A. infectives roughly 2.6% that experienced in the U.S.A. could sustain a Canadian epidemic at a comparative incidence ratio of around 4 to 1, U.S. to Canadian. (If someone would conjecture that it is rather the Canadian infectives that are causing the epidemic in the United States, that would require that the activity rate of Canadian infectives with U.S. susceptibles must be $1/.026 = 38.5$ times that of Canadian infectives with Canadian susceptibles.) If this activity would double to 5.2%, then the Canadian total infectives would double, but the rate $(1/x_{Can})dx_{Can}/dt$ would still grow at rate β_t.

Similarly, we can show that

$$\begin{aligned}
\eta_{Fr,t} &= .076\beta_t \\
\eta_{U.K.,t} &= .024\beta_t \\
\eta_{Dk,t} &= .0034\beta_t \\
\eta_{NL,t} &= .0075\beta_t.
\end{aligned}$$

12.4   America and the First World Epidemic

We have observed some surprises and tried to come up with plausible explanations for those surprises. The relative incidence of AIDS for various First World countries when compared to that of the United States appears, for each country, to be relatively constant over time and this incidence appears to be roughly the same for cumulative ratios and for ratios of new cases. The rate of growth for AIDS, β_t, changes year by year, but it seems to be nearly the same for all the First World countries considered (Figure 12.6), including the United States. The bathhouse phenomenon that I have repeatedly argued as essential for the maintenance of a stand-alone epidemic of AIDs in a First World country is generally not present in First World countries other than the United States. Therefore, a stand-alone epidemic there is not likely. Yet AIDS has a continuing small (compared to that of the USA), though significant, presence in First World countries other than the United States. The new case (piecewise exponential) rate there tracks that of the United States rather closely, country by country. We have shown that a model where a Poissonian term for "travel" from and to the USA is dominant does show one way in which one can explain the surprises. In 1984 [3], I pointed out that the American gay community was made unsafe by the presence of a small subpopulation which visited the bathhouses, even though the large majority of gays, as individuals, might not frequent these establishments. The present work gives some indication that the high AIDS incidence in the United States should be a matter of concern to other First World countries as long as travel to and from the United States continues at the brisk rates we have seen since the early 1980s.

Nothing could have been easier than U.S. officials closing down the bathhouses in the early 1980's, an action taken quickly by the French and by other First World countries who had bathhouses or bathhouse surrogates. That the American officials did not take similar steps has caused the infection of around 1 million Americans with a fatal disease. There is some acknowledgment within the leadership of the American gay community of the effect of the small subgroup of high sexual activity in pushing AIDS from endemic to epidemic in the United States [1].

Now, we seem to have strong EDA indications concerning the essential role the United States plays in maintaining AIDS as an "epidemic" in other First World countries. If it becomes clear that I am right, this will be a problem difficult for Europeans to address. Are a few tens of thousands of lives of sufficient value that blood tests be required for visas for persons, particularly Americans and citizens of other high-HIV-incidence countries, entering the European Union? What warnings should be given to the nationals of one's own country who contemplate visits to a country with a high incidence of HIV infection? Does the failure of a First World country, such as the United States, to take reasonable steps to stifle an epidemic, creating a pool of infectives which drives the epidemic in other countries, produce liabilities for the offending First World country?

12.5   Modeling The Bathhouse Effect

First, we should note that a distinguishing epidemiological characteristic of AIDS as opposed to the more classical venereal diseases is its low infectivity. As a "back of the envelope" example, let us suppose that we are dealing with a very infective disease, one in which one contact changes a susceptible to an infective. Then let us suppose that we have an infective who is going to engage in five contacts. What number of susceptibles (assuming equal mixing) will give the highest expected number of conversions of susceptibles to infectives? Note that if the number of susceptibles is small, the expectation will be lessened by the "overkill effect," (i.e., there is the danger that some of the contacts will be "wasted" by being applied to an individual already infected by one of the other five contacts). Clearly, here, the optimal value for the numerosity N of the susceptible pool is infinity, for then the expected number of conversions from susceptible to infective $E(N = \infty)$ is five.

Now, let us change the situation to one where two contacts, rather than one, are required to change a susceptible to an infective. We will still assume a total of five contacts. Clearly, if $N=1$, then the expected number of conversions $E=1$. Clearly, there has been wastage due to overkill. Next, let us assume the number of susceptibles has grown to $N=2$. Then the probability of two new infectives is given by

$$P(2|N = 2) = \sum_{j=2}^{3} \binom{5}{j} \left(\frac{1}{2}\right)^5 = \frac{20}{32}. \qquad (12.15)$$

The probability of only one new infective is $1 - P(2|N = 2)$. Thus the expected number of new infectives is

$$E(\mathcal{I}|N=2) = 2\frac{20}{32} + 1\frac{12}{32} = 1.625. \tag{12.16}$$

Now, when there are $N = 3$ susceptibles, the contact configurations leading to two new infectives are of the sort (2,2,1) and (3,2,0). All other configurations will produce only one new infective. So the probability of two new infectives is given by

$$P(2|N=3) = \binom{3}{1} \frac{5!}{2!2!1!} \left(\frac{1}{3}\right)^5 + \binom{3}{1}\binom{2}{1}\frac{5!}{3!2!}\left(\frac{1}{3}\right)^5 = \frac{150}{243} \tag{12.17}$$

and the expected number of new infectives is given by

$$E(\mathcal{I}|N=3) = 2\frac{150}{243} + 1\frac{93}{243} = 1.617. \tag{12.18}$$

Going further, we find that $E(\mathcal{I}|N=4) = 1.469$ and $E(\mathcal{I}|N=5) = 1.314$. For very large N, $E(\mathcal{I})$ is of order $1/N$. Apparently, for the situation where there is a total of five contacts, the value of the number in the susceptible pool that maximizes the total number of new infectives from the one original infective is $N = 2$, not ∞. Obviously, we are oversimplifying, since we stop after only the contacts of the original infective. The situation is much more complicated here, since an epidemic is created by the new infectives infecting others, and so on. Then there is the matter of a distribution of the number of contacts required to give the disease; and so on. We shall avoid unnecessary branching process modeling by going deterministic. The argument above is given only to present an intuitive feel as to the facilitating potential of a high contact core to driving a disease over the threshold of sustainability.

Now for AIDS, the average number of anal intercourses required to break down the immune system sufficiently to cause the person ultimately to get AIDS is much larger than two. The obvious implication is that a great facilitator for the epidemic being sustained is the presence of a subpopulation of susceptibles whose members have many contacts. In the "back of the envelope" example above, we note that even if the total number of contacts was precisely five, it was best to concentrate the contacts in a small pool of susceptibles. In other words, if the total number of contacts is fixed at some level, it is best to start the epidemic by concentrating the contacts amongst a small population. Perhaps the analogy to starting a fire, not by dropping a match onto a pile of logs, but rather onto some kindling beneath the logs, is helpful.

In the case of modeling the AIDS epidemic (and most epidemics), very little is gained by using a stochastic model [9]. When dealing with large populations, generally speaking, we can deal with most of the relevant aspects by looking at deterministic models. However, to develop even a deterministic model for AIDS could, quite easily, involve hundreds of variables. At this time, we have neither the data nor the understanding to justify such a model for AIDS. Then, one could argue, as some have, that we should disdain modeling altogether until such time as we have sufficient information. But to give up modeling until "all the facts are in" would surely push us past the time where our analyses would be largely irrelevant. There is every reason to hope that at some future time we will have a vaccine and/or a treatment for AIDS. This was the case, for example, with polio. Do we now have the definitive models for polio? The time for developing models of an epidemic is during the period when they might be of some use. The model below was first presented in 1984 [3]. (A more general version was presented in 1989 [6], but the insights are the same when the simpler model is used.) We begin with a classical contact formulation:

$$P(\text{transmission from infective in } [t, t + \Delta t)) = k\alpha\Delta t\frac{X}{X+Y}, \quad (12.19)$$

where

$k =$ number of contacts per month;
$\alpha =$ probability of contact causing AIDS;
$X =$ number of susceptibles;
$Y =$ number of infectives.

We then seek the expected total increase in the infective population during $[t, t + \Delta t)$ by multiplying the above by the total number of infectives:

$$\Delta E(Y) = YP(\text{transmission in } [t, t + \Delta t) \approx \Delta Y. \quad (12.20)$$

Letting $\Delta t \to 0$, we have

$$\frac{dY}{dt} = \frac{k\alpha XY}{X+Y}, \quad (12.21)$$

$$\frac{dX}{dt} = -\frac{k\alpha XY}{X+Y}. \quad (12.22)$$

There are other factors that must be added to the model, such as immigration into the susceptible population, λ, and emigration, μ, from both the susceptible and infective populations, as well as a factor γ to allow for marginal increase in the emigration from the infective population due to sickness and death. Thus, we have the improved differential equation model,

$$\frac{dY}{dt} = \frac{k\alpha XY}{X+Y} - (\gamma + \mu)Y, \quad (12.23)$$

$$\frac{dX}{dt} = -\frac{k\alpha XY}{X+Y} + \lambda - \mu X. \quad (12.24)$$

Let us now proceed away from the situation where it is assumed that all persons in the gay population have equal contact rates to one where there are two populations: the majority, less sexually active, but with a minority (e.g., bathhouse visitors) with greater activity than that of the majority. In the following, we use the subscript "1" to denote the majority, less sexually active portion of the target (gay) population, and the subscript "2" to denote the minority, sexually very active portion (the part that engages in high-frequency anonymous anal intercourse, typically at bathhouses). The more active population will be taken to have a contact rate τ times that of the rate k of the majority portion of the target population. The fraction of the more sexually active population will be taken to be p.

$$\frac{dY_1}{dt} = \frac{k\alpha X_1(Y_1 + \tau Y_2)}{X_1 + Y_1 + \tau(Y_2 + X_2)} - (\gamma + \mu)Y_1;$$

$$\frac{dY_2}{dt} = \frac{k\alpha\tau X_2(Y_1 + \tau Y_2)}{X_1 + Y_1 + \tau(Y_2 + X_2)} - (\gamma + \mu)Y_2;$$

$$\frac{dX_1}{dt} = -\frac{k\alpha X_1(Y_1 + \tau Y_2)}{X_1 + Y_1 + \tau(Y_2 + X_2)} + (1 - p)\lambda - \mu X_1;$$

$$\frac{dX_2}{dt} = -\frac{k\alpha\tau X_2(Y_1 + \tau Y_2)}{X_1 + Y_1 + \tau(Y_2 + X_2)} + p\lambda - \mu X_2, \qquad (12.25)$$

where

k = number of contacts per month;
α = probability of contact causing AIDS;
λ = immigration rate into sexually active gay population;
μ = emigration rate from sexually active gay population;
γ = marginal emigration rate from sexually active gay
    population due to sickness and death;
X = number of susceptibles;
Y = number of infectives.

In [3], it was noted that if we started with 1000 infectives in a gay population with $k\alpha = .05$, susceptible population of 3,000,000, the best guesses then available [$\mu = 1/(15 \times 12) = .00556$, $\gamma = 0.1$, $\lambda = 16,666$] for the other parameters, and $\tau = 1$, the disease advanced as shown in Table 12.3.

Table 12.3. Extrapolated AIDS Cases.		
$k\alpha=.05$; $\tau = 1$		
Year	Cumulative Deaths	Fraction Infective
1	1751	.00034
2	2650	.00018
3	3112	.00009
4	3349	.00005
5	3571	.00002
10	3594	.000001

Next, a situation was considered where the overall contact numerosity was the same as in Table 12.3, but there are two activity levels with the more sexually active subpopulation (of size 10%) having contact rates 16 times those of the less active population as shown in Table 12.4.

Table 12.4. Extrapolated AIDS Cases.		
$k\alpha = .02$; $\tau = 16$; $p = .10$		
Year	Cumulative Deaths	Fraction Infective
1	2,184	.0007
2	6,536	.0020
3	20,583	.0067
4	64,157	.0197
5	170,030	.0421
10	855,839	.0229
15	1,056,571	.0122
20	1,269,362	.0182

Although the overall average contact rate in Tables 12.3 and 12.4 is the same $(k\alpha)_{overall} = .05$, the situation is dramatically different in the two cases. Here, it seemed, was a *prima facie* explanation as to how AIDS was pushed over the threshold to a full-blown epidemic in the United States: a sexually very active subpopulation (i.e., the customers of the gay bathhouses).

We note that even with a simplified model such as that presented here, we appear to be hopelessly overparameterized. There is little chance that we shall have reliable estimates of all of $k, \alpha, \gamma, \mu, \lambda, p$, and τ. One of the techniques sometimes available to the modeler is to express the problem in such a form that most of the parameters will cancel. For the present case we will attempt to determine the $k\alpha$ value necessary to sustain the epidemic for the heterogeneous case when the number of infectives is very small. For $Y_1 = Y_2 = 0$ the equilibrium values for X_1 and X_2 are $(1-p)(\lambda/\mu)$ and $p(\lambda/\mu)$, respectively. Expanding the right-hand sides of (12.25) in a Maclaurin series, we have (using lower case symbols for the perturbations)

$$\frac{dy_1}{dt} = \left[\frac{k\alpha(1-p)}{1-p+\tau p} - (\gamma+\mu)\right] y_1 + \frac{k\alpha(1-p)\tau}{1-p+\tau p} y_2$$

$$\frac{dy_2}{dt} = \frac{k\alpha\tau p}{1-p+\tau p} y_1 + \left[\frac{k\alpha\tau^2 p}{1-p+\tau p} - (\gamma+\mu)\right] y_2.$$

Summing then gives

$$\frac{dy_1}{dt} + \frac{dy_2}{dt} = [k\alpha - (\gamma+\mu)] y_1 + [k\alpha\tau - (\gamma+\mu)] y_2. \qquad (12.26)$$

Assuming a neglible number of initial infectives, for the early stages of the epidemic we have

$$y_1 = \frac{(1-p)}{p\tau} y_2. \qquad (12.27)$$

Substituting in (12.26) shows that for the epidemic to be sustained, we must have

$$k\alpha > \frac{(1-p+\tau p)}{1-p+\tau^2 p}(\gamma+\mu). \tag{12.28}$$

Accordingly we define the *heterogeneous threshold* via

$$k^*\alpha = \frac{(1-p+\tau p)}{1-p+\tau^2 p}(\gamma+\mu). \tag{12.29}$$

Now, in the homogeneous contact case (i.e., $\tau = 1$), we note that for the epidemic to be sustained we require

$$k_H\alpha > (\gamma+\mu). \tag{12.30}$$

For the heterogeneous contact case with k^*, the average contact rate is given by

$$k_{ave}\alpha \; = \; p\tau(k^*\alpha) + (1-p)(k^*\alpha) \tag{12.31}$$

$$= \; \frac{(1-p+\tau p)^2}{1-p+\tau^2 p}(\gamma+\mu). \tag{12.32}$$

So, dividing the sustaining $k_H\alpha$ by the sustaining value for the heterogeneous contact rate, we have

$$Q = \frac{1-p+\tau^2 p}{(1-p+\tau p)^2}. \tag{12.33}$$

We note that we have been able here to reduce the parameters necessary for consideration from seven to two. This is fairly typical for model-based approaches: the dimensionality of the parameter space may be reducible in answering specific questions. It is shown elsewhere [6] that the addition of time delay effects between infection and infectiousness to the model still yields precisely the enhancement factor shown in Figure 12.9. In that figure, we note a plot of this "bathhouse enhancement factor." Note that the addition of the bathhouses to the transmission picture had roughly the same effect as if all the members of the target population had doubled their contact rate. Remember that the picture has been corrected to discount any increase in the overall contact rate which occurred as a result of the addition of the bathhouses. In other words, the enhancement factor is totally due to heterogeneity. Here τ is the activity multiplier for the sexually very active gay subpopulation and p is the proportion of the total sexually active gay population which it constitutes. We note that *for the same total number of sexual contacts across the gay population*, the presence of a small high activity subpopulation can have roughly the same effect as if the entire gay population had doubled its sexual activity. It is this heterogeneity effect which I have maintained (since 1984) to be the cause of AIDS getting over the threshold of sustainability in the United States.

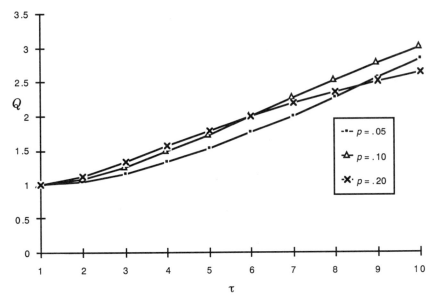

Figure 12.9. Effect of High Activity Subpopulation.

For a fixed value of τ, we have that Q is maximized when

$$p = \frac{1}{1+\tau}. \qquad (12.34)$$

For this value of p, we have

$$Q = \frac{(1+\tau)^2}{4\tau}. \qquad (12.35)$$

12.6   Heterogeneity Effects in the Mature Epidemic

The AIDS epidemic in the United States has long passed the point where perturbation threshold approximations can be used assuming a small number of infectives. One might well ask whether bathhouse closings at this late date would have any benefit.

To deal with this question, we unfortunately lose our ability to disregard five of the seven parameters and must content ourselves with picking reasonable values for those parameters. A detailed analysis is given in [7]. Here we shall content ourselves with looking at the case where the contact rate before the possible bathhouse closings is given by

$$(k\alpha)_{overall} = (1 - p + \tau p)(\gamma + \mu). \qquad (12.36)$$

Furthermore, we shall take $\mu = 1/(180 \text{ months})$ and $\lambda = 16,666$ per month. (We are assuming a target population, absent the epidemic, of roughly 3,000,000.) For a given fraction of infectives in the target population of π, we want to determine the ratio of contact rates causing elimination of the epidemic for the closings case divided by that without closings. Such a picture is given in Figure 12.10. It would appear that as long as the proportion of infectives is no greater than 40% of the target population, there would be a benefit from bathhouse closings. Unfortunately, in most large cities in the United States, this infectivity rate may well have been exceeded already.

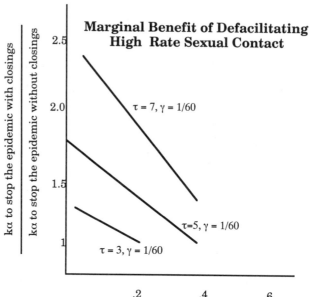

Figure 12.10. Bathhouse Closing Effects.

12.7   Conclusions

This chapter gives us an insight into an evolving epidemic. When the first paper [3] was presented on the bathhouses as the likely cause of AIDS being driven over the epidemiological threshold, the data was rather limited. The total number of AIDS cases in the United States at the time of the preparation of the paper was only a few thousand, and the number of cases in Europe was an order of magnitude less than that. If a timely analysis which might be useful for the formulation of policy was to take place, then one had to deal with the "data" at hand. The fact was that a new epidemic was clearly experiencing classical exponential growth. There seemed to be no records of this epidemic before, and, historically, cases of individuals with AIDS-like symptoms were so rare that one learned of them anecdotally from pathologists. At the time (and, unfortunately, still) there was the concept that AIDS was caused by some new virus. But the maxim of the virologist is that, just as no new species of animals are being formed, no new viruses are either. The practical approach of the epidemiologist is to look for sociological causes for the outbreak of an epidemic (e.g., a contaminated source of drinking water, the arrival of persons from places where the epidemic is already present, etc.). The idea of the change being a new acceptance of homosexuality in the United States did not square with the time line. Homosexuality was accepted already by 1970 to the degree where large gay communities were thriving from San Francisco and Los Angeles to New York and Miami and Houston. Yet the epidemic did not get going until the early 1980s. One needed to find another potential cause. The bathhouses, which only achieved civil acceptance in America by the late 1970s, seemed to be a possibility. All this is typical of the way one gets "data" in an epidemic. The goal should be to stop the epidemic early on, when data is skimpy. The benefit of the doubt must be given to the greater society when policy is being formed. And Thompson [3–6] was able to show how a small highly active gay subset could have the same effect as if the entire American gay population had doubled its average activity.

At the time of the publication of [3] in 1984, there was little motion on the part of U.S. public health officials to shutting down the bathhouses and their surrogates. (In fact, they are still not closed.) It was also clear that the Europeans were shutting them down with essentially no resistance. The results of these varying policies now seem rather clear, as we have seen in the first part of this chapter. The United States AIDS epidemic rages as a stand-alone with expected fatalities in excess of 1 million. The Europeans have rates which are an order of magnitude less than those in the United States. Moreover, as we have seen, their epidemics are not stand-alone, but rather due to contacts with the huge pool of American infectives. A failure to build public policy based on facts will cost the United States more dead citizens than all the wars of the twentieth century. And there is the added humiliation that the USA is essentially a Typhoid Mary country, causing AIDS epidemics in other First World countries. The tragedy of building policy based on political expediency rather than reality is very clear in the case of AIDS. And, like a Greek tragedy, it proceeds painfully and slowly towards an impending doom.

Throughout this book, it has been noted that modern computing has had the tendency to drive us from models. Digital computers are not analog devices, and many people have allowed that fact to form their ways of thinking. But the hallmark of Western thought since the Greeks has been analogical reasoning. The mismanagement of the United States AIDS epidemic is one example of the danger of attempting to look at data without building models to explain them and give us insights as to how underlying systems might be changed for the better.

The union of data, models, and high-speed computing holds out the promise of a powerful new paradigm for posing and solving problems. The digital computer in 50 years has advanced to undreamed-of speeds. But the maturity of the users of computers has lagged far behind the hardware. Our ability to get full use from computers will only be realized when we are willing to use them as a means of enhancing our modeling abilities rather than as a replacement for them.

Problems

12.1. Using the data in Table 12.1, one can obtain estimates for the piecewise growth rates from equation (12.2) as shown in Table 12.5. Construct a bootstrap test to test the hypothesis that, year by year, the kinetic constant is the same for the United States as for the other countries shown.

Table 12.5. Estimates of Kinetic Constants					
Year	U.S.A.	U.K.	Canada	Denmark	France	Netherlands
1985	0.518	0.597	0.584	0.513	0.611	0.540
1986	0.459	0.535	0.498	0.482	0.569	0.523
1987	0.413	0.434	0.428	0.411	0.504	0.482
1988	0.347	0.367	0.341	0.342	0.406	0.393
1989	0.290	0.304	0.289	0.320	0.336	0.321
1990	0.251	0.265	0.230	0.266	0.263	0.256
1991	0.229	0.224	0.201	0.221	0.232	0.214
1992	0.232	0.202	0.182	0.180	0.205	0.196
1993	0.187	0.184	0.155	0.171	0.179	0.153
1994	0.143	0.158	0.131	0.144	0.163	0.138
1995	0.113	0.127	0.101	0.116	0.123	0.120

12.2. The combinatorics of finding the expected number of infectives created in the early days of an epidemic [as in equation (12.18)] can quickly grow tedious. Moreover, it is very easy to make mistakes. Resampling gives us an easy way out. If there are n contacts to be spread among N individuals in a short period of time (say, the time of infectivity of the infectives), we may repeatedly take integer samples from 1 to N and count the fraction of times integers are repeated n or more times. Using this approach, if a total of 10 contacts are to be made, find the size of the susceptible pool which gives the largest number of expected infections, given that at least three contacts are required to convert a susceptible into an infective.

12.3. In Table 12.6 we show the ratio of the cumulative incidence of AIDS per 100,000 population for the United States divided by that for the United Kingdom, Canada, Denmark, France, and the Netherlands. Construct a resampling based test of the hypothesis that these ratios are constant over the 10 year period considered.

Table 12.6. Ratios of U.S.A. AIDS Incidences to Those of Other Countries				
Year	U.K.	Canada	Denmark	France	Netherlands
1985	12.427	3.735	10.693	6.092	5.254
1986	10.695	3.468	9.432	5.831	4.193
1987	10.300	3.379	8.318	5.842	3.541
1988	9.982	3.405	7.727	5.888	3.219
1989	9.784	3.408	7.389	5.635	3.010
1990	9.597	3.505	7.346	5.521	2.960
1991	9.669	3.636	7.489	5.581	2.949
1992	10.048	3.871	7.833	5.954	3.050
1993	10.088	4.023	8.163	6.075	3.083
1994	9.904	4.080	8.208	6.067	3.012
1995	9.744	4.138	8.140	6.048	2.977

12.4. The assumption of a sexually very active subpopulation is, of course, not the only way to bring AIDS to epidemic levels. Redo Table 12.3 but make $\gamma = .01$. This scenario has increased the sexually active period of an AIDS infective from 10 months to 100 months.

12.5. Computers have become so fast, storage so plentiful, that we are tempted to dispense with differential equation aggregates and work directly with the underlying axioms. Such an approach was suggested in Chapter 3. Let us consider one not quite atomistic approach. Create a population of 300 susceptibles, 30 of whom have an activity level τ times that of the dominant population. Suppose that one (high activity) infective is introduced into the population. Keep track of all the members of the population as susceptible individuals S, "retired" susceptible individuals R, infective individuals $\mathcal{I}$, and dead individuals D. See what τ needs to be to sustain the epidemic with high probablity. At the beginning of the time interval $[t, t + \Delta t)$,

$$P(\text{new susceptible appears in } [t, t + \Delta t]) = \lambda \Delta t.$$

If such a person appears, we add him to the number of susceptibles, according to the proportion of .10 for high activity, .90 for low activity.

Then a susceptible may, for whatever reason, remove himself from the pool of risk.

$$P(\text{susceptible } J, \text{ "retires" in } [t, t + \Delta t)) = \mu \Delta t.$$

If this happens, we remove him from the infective pool and add him to the retired pool.

Next, an infective may die (or be so sick as to be inactive):

$$P(\text{an infective dies in } [t, t + \Delta t)) = \gamma \Delta t.$$

If this happens, we remove him from the pool of infectives and add him to the list of the dead.

Then, for each susceptible person,

$$P(\text{low-activity susceptible, converts to infective in } [t, t + \Delta t))$$
$$= \frac{k\alpha\Delta t X_1(Y_1 + \tau Y_2)}{X_1 + Y_1 + \tau(Y_2 + X_2)}.$$

If such a change is made, we add the individual to the pool of low-activity infectives removing him from the pool of low-activity susceptibles. Similarly,

$$P(\text{high-activity susceptible, converts to infective in } [t, t + \Delta t))$$
$$= \frac{\tau k\alpha\Delta t X_1(Y_1 + \tau Y_2)}{X_1 + Y_1 + \tau(Y_2 + X_2)}.$$

If such a conversion takes place, we remove the person from the pool of high-activity susceptibles, adding him to the pool of high-activity infectives. where

τ = multiple of number of contacts of low-activity
     population for high-activity population;
k = number of contacts per month;
α = probability of contact causing AIDS;
λ = immigration rate into sexually active gay population;
μ = emigration rate from sexually active gay population;
γ = marginal emigration rate from sexually active
     gay population due to sickness and death;
X_1 = number of low-activity susceptibles;
X_2 = number of high-activity susceptibles;
Y_1 = number of low-activity infectives;
Y_2 = number of high-activity infectives.

This problem may well indicate the reason that "higher-order" languages are frequently not the choice for nontrivial simulations, which are generally DO-LOOP intensive. The running time for this program in FORTRAN or C is a tiny fraction of that required when the program is written in MATLAB or SPLUS. There is, naturally, no particular reason why this need be true. There is no reason why DO-LOOPS cannot be accommodated if only the compiler be written to do so.

12.6. There are many processes of the empirical birth-and-death variety related to those for epidemics. For example, there is the whole topic of simulating warfare. Suppose we have two sides, the Red and the Blue. Then we may ([4], pp. 55−71), if there are n subforces of Red, and m subforces of Blue, write down the heterogeneous force Lanchester equations

$$\frac{du_j}{dt} = -\sum_{i=1}^{n} k_{ij}c_{1ij}v_i$$

$$\frac{dv_i}{dt} = -\sum_{i=1}^{n} l_{ji}c_{2ji}u_j,$$

where k_{ij} represents the allocation (between 0 and 1) such that $\sum_{j=1}^{m} k_{ij} \leq 1$ of the ith Red subforce's firepower against the jth Blue subforce. Also, c_{1ij} represents the attrition coefficient of the ith Red subforce against the jth Blue subforce; and similarly for l_{ji} and c_{2ji}. Write down the stochastic laws that one might use to simulate this system at the unit level (e.g., one company of Red tanks against two companies of Blue infantry). Such procedures for combat attrition were used since von Reiswitz introduced them in 1820 (with dice tosses and patient officers sitting around game boards). Interestingly, such games can easily be computerized, and their concordance with historical reality is excellent.

References

[1] Rotello, G. (1997). *Sexual Ecology: AIDS and the Destiny of Gay Men.* New York: Dutton, 85−89.

[2] Shilts, R. (1987). *And the Band Played On: Politics, People, and the AIDS Epidemic.* New York: St. Martin's Press, 144.

[3] Thompson, J.R. (1984). "Deterministic versus stochastic modeling in neoplasia," *Proceedings of the 1984 Computer Simulation Conference.* New York: North-Holland, 822−825.

[4] Thompson, J.R. (1989). *Empirical Model Building.* New York: John Wiley & Sons, 79−91.

[5] Thompson, J.R. and Tapia, R.A. (1990). *Nonparametric Function Estimation, Modeling and Simulation.* Philadelphia: SIAM, 233−243.

[6] Thompson, J.R. (1989). "AIDS: the mismanagement of an epidemic," *Comput. Math. Applic.,* **18**, 965−972.

[7] Thompson, J.R. and Go, K. (1991). "AIDS: modeling the mature epidemic in the American gay community," in *Mathematical Population Dynamics,* O. Arino, D. Axelrod and M. Kimmel, eds. New York: Marcel Dekker, 371−382.

[8] Thompson, J.R.(1998). "The United States AIDS Epidemic in First World Context," in *Advances in Mathematical Population Dynamics: Molecules, Cells and Man*, O. Arino, D. Axelrod, and M. Kimmel, eds. Singapore: World Scientific Publishing Company, 345–354.

[9] West, R.W. (1994). *Modeling the Potential Impact of HIV on the Spread of Tuberculosis in the United States*. Doctoral Dissertation. Houston: Rice University

[10] West, R.W. and Thompson, J.R. (1998). "Modeling the impact of HIV on the spread of tuberculosis in the United States," *Math. Biosci.*, **143**, 35–60.

[11] West, R.W. and Thompson, J.R. (1998). "Models for the Simple Epidemic," *Math. Biosci.*, **141**, 29–39.

Appendix: Statistical Tables

1. Table of the Normal Distribution

				Values of $\frac{1}{\sqrt{2\pi}} \int_{-\infty}^{z} e^{-t^2/2} dt$					
z	.0	.01	.02	.03	.04	.05	.06	.07	.08	.09
.0	.5	.5040	.5080	.5120	.5160	.5199	.5239	.5279	.5319	.5359
.1	.5398	.5438	.5478	.5517	.5557	.5596	.5636	.5675	.5714	.5753
.2	.5793	.5832	.5871	.5910	.5948	.5987	.6026	.6064	.6103	.6141
.3	.6179	.6217	.6255	.6293	.6331	.6368	.6406	.6443	.6480	.6517
.4	.6554	.6591	.6628	.6664	.6700	.6736	.6772	.6808	.6844	.6879
.5	.6915	.6950	.6985	.7019	.7054	.7088	.7123	.7157	.7190	.7224
.6	.7257	.7291	.7324	.7357	.7389	.7422	.7454	.7486	.7517	.7549
.7	.7580	.7611	.7642	.7673	.7704	.7734	.7764	.7794	.7823	.7852
.8	.7881	.7910	.7939	.7967	.7995	.8023	.8051	.8078	.8106	.8133
.9	.8159	.8186	.8212	.8238	.8264	.8289	.8315	.8340	.8365	.8389
1.0	.8413	.8438	.8461	.8485	.8508	.8531	.8554	.8577	.8599	.8621
1.1	.8643	.8665	.8686	.8708	.8729	.8749	.8770	.8790	.8810	.8830
1.2	.8849	.8869	.8888	.8907	.8925	.8944	.8962	.8980	.8997	.9015
1.3	.9032	.9049	.9066	.9082	.9099	.9115	.9131	.9147	.9162	.9177
1.4	.9192	.9207	.9222	.9236	.9251	.9265	.9279	.9292	.9306	.9319
1.5	.9332	.9345	.9357	.9370	.9382	.9394	.9406	.9418	.9429	.9441
1.6	.9452	.9463	.9474	.9484	.9495	.9505	.9515	.9525	.9535	.9545
1.7	.9554	.9564	.9573	.9582	.9591	.9599	.9608	.9616	.9625	.9633
1.8	.9641	.9649	.9656	.9664	.9671	.9678	.9686	.9693	.9699	.9706
1.9	.9713	.9719	.9726	.9732	.9738	.9744	.9750	.9756	.9761	.9767
2.0	.9772	.9778	.9783	.9788	.9793	.9798	.9803	.9808	.9812	.9817
2.1	.9821	.9826	.9830	.9834	.9838	.9842	.9846	.9850	.9854	.9857
2.2	.9861	.9864	.9868	.9871	.9875	.9878	.9881	.9884	.9887	.9890
2.3	.9893	.9896	.9898	.9901	.9904	.9906	.9909	.9911	.9913	.9916
2.4	.9918	.9920	.9922	.9925	.9927	.9929	.9931	.9932	.9934	.9936
2.5	.9938	.9940	.9941	.9943	.9945	.9946	.9948	.9949	.9951	.9952
2.6	.9953	.9955	.9956	.9957	.9959	.9960	.9961	.9962	.9963	.9964
2.7	.9965	.9966	.9967	.9968	.9969	.9970	.9971	.9972	.9973	.9974
2.8	.9974	.9975	.9976	.9977	.9977	.9978	.9979	.9979	.9980	.9981
2.9	.9981	.9982	.9982	.9983	.9984	.9984	.9985	.9985	.9986	.9986
3.0	.9987	.9987	.9987	.9988	.9988	.9989	.9989	.9989	.9990	.9990
3.1	.9990	.9991	.9991	.9991	.9992	.9992	.9992	.9992	.9993	.9993
3.2	.9993	.9993	.9994	.9994	.9994	.9994	.9994	.9995	.9995	.9995
3.3	.9995	.9995	.9995	.9996	.9996	.9996	.9996	.9996	.9996	.9997
3.4	.9997	.9997	.9997	.9997	.9997	.9997	.9997	.9997	.9997	.9998
3.5	.9998	.9998	.9998	.9998	.9998	.9998	.9998	.9998	.9998	.9998
3.6	.9998	.9998	.9999	.9999	.9999	.9999	.9999	.9999	.9999	.9999
3.7	.9999	.9999	.9999	.9999	.9999	.9999	.9999	.9999	.9999	.9999
3.8	.9999	.9999	.9999	.9999	.9999	.9999	.9999	.9999	.9999	.9999

2. Table of the χ^2 Distribution

Critical Values of $P = \frac{1}{2^{\nu/2}\Gamma(\nu/2)} \int_0^{\chi^2} x^{\nu/2-1} e^{-x/2} dx$							
$\nu \backslash P$	0.100	0.500	0.750	0.900	0.950	0.975	0.990
1	0.016	0.455	1.323	2.706	3.841	5.024	6.635
2	0.211	1.386	2.773	4.605	5.991	7.378	9.210
3	0.584	2.366	4.108	6.251	7.815	9.348	11.345
4	1.064	3.357	5.385	7.779	9.488	11.143	13.277
5	1.610	4.351	6.626	9.236	11.070	12.833	15.086
6	2.204	5.348	7.841	10.645	12.592	14.449	16.812
7	2.833	6.346	9.037	12.017	14.067	16.013	18.475
8	3.490	7.344	10.219	13.362	15.507	17.535	20.090
9	4.168	8.343	11.389	14.684	16.919	19.023	21.666
10	4.865	9.342	12.549	15.987	18.307	20.483	23.209
11	5.578	10.341	13.701	17.275	19.675	21.920	24.725
12	6.304	11.340	14.845	18.549	21.026	23.337	26.217
13	7.042	12.340	15.984	19.812	22.362	24.736	27.688
14	7.790	13.339	17.117	21.064	23.685	26.119	29.141
15	8.547	14.339	18.245	22.307	24.996	27.488	30.578
16	9.312	15.338	19.369	23.542	26.296	28.845	32.000
17	10.085	16.338	20.489	24.769	27.587	30.191	33.409
18	10.865	17.338	21.605	25.989	28.869	31.526	34.805
19	11.651	18.338	22.718	27.204	30.144	32.852	36.191
20	12.443	19.337	23.828	28.412	31.410	34.170	37.566
21	13.240	20.337	24.935	29.615	32.671	35.479	38.932
22	14.041	21.337	26.039	30.813	33.924	36.781	40.289
23	14.848	22.337	27.141	32.007	35.172	38.076	41.638
24	15.659	23.337	28.241	33.196	36.415	39.364	42.980
25	16.473	24.337	29.339	34.382	37.652	40.646	44.314
26	17.292	25.336	30.435	35.563	38.885	41.923	45.642
27	18.114	26.336	31.528	36.741	40.113	43.195	46.963
28	18.939	27.336	32.620	37.916	41.337	44.461	48.278
29	19.768	28.336	33.711	39.087	42.557	45.722	49.588
30	20.599	29.336	34.800	40.256	43.773	46.979	50.892

3. Table of the Student t Distribution

Critical Values of $P = \frac{\Gamma[(\nu+1)/2]}{\Gamma(\nu/2)\sqrt{\pi\nu}} \int_{-\infty}^{t}(1+\frac{w^2}{\nu})^{(\nu+1)/2}dw$					
$\nu\backslash P$	0.750	0.900	0.950	0.975	0.990
1	1.0000	3.0777	6.3138	12.7062	31.8205
2	0.8165	1.8856	2.9200	4.3027	6.9646
3	0.7649	1.6377	2.3534	3.1824	4.5407
4	0.7407	1.5332	2.1318	2.7764	3.7469
5	0.7267	1.4759	2.0150	2.5706	3.3649
6	0.7176	1.4398	1.9432	2.4469	3.1427
7	0.7111	1.4149	1.8946	2.3646	2.9980
8	0.7064	1.3968	1.8595	2.3060	2.8965
9	0.7027	1.3830	1.8331	2.2622	2.8214
10	0.6998	1.3722	1.8125	2.2281	2.7638
11	0.6974	1.3634	1.7959	2.2010	2.7181
12	0.6955	1.3562	1.7823	2.1788	2.6810
13	0.6938	1.3502	1.7709	2.1604	2.6503
14	0.6924	1.3450	1.7613	2.1448	2.6245
15	0.6912	1.3406	1.7531	2.1314	2.6025
16	0.6901	1.3368	1.7459	2.1199	2.5835
17	0.6892	1.3334	1.7396	2.1098	2.5669
18	0.6884	1.3304	1.7341	2.1009	2.5524
19	0.6876	1.3277	1.7291	2.0930	2.5395
20	0.6870	1.3253	1.7247	2.0860	2.5280
21	0.6864	1.3232	1.7207	2.0796	2.5176
22	0.6858	1.3212	1.7171	2.0739	2.5083
23	0.6853	1.3195	1.7139	2.0687	2.4999
24	0.6848	1.3178	1.7109	2.0639	2.4922
25	0.6844	1.3163	1.7081	2.0595	2.4851
26	0.6840	1.3150	1.7056	2.0555	2.4786
27	0.6837	1.3137	1.7033	2.0518	2.4727
28	0.6834	1.3125	1.7011	2.0484	2.4671
29	0.6830	1.3114	1.6991	2.0452	2.4620
30	0.6828	1.3104	1.6973	2.0423	2.4573
40	0.6807	1.3031	1.6839	2.0211	2.4233
60	0.6786	1.2958	1.6706	2.0003	2.3901
120	0.6765	1.2886	1.6577	1.9799	2.3578
∞	0.6745	1.2816	1.6450	1.9602	2.3267

4a. Table of the F Distribution with $\alpha = .05$

Critical Values of F Test when $\alpha = .05$

$\nu_2 \backslash \nu_1$	1	2	3	4	5	6	8	10	20	30	60	∞
1	161.4476	199.5000	215.7073	224.5832	230.1619	233.9860	238.8827	241.8817	248.0131	250.0951	252.1957	254.3017
2	18.5128	19	19.1643	19.2468	19.2964	19.3295	19.3710	19.3959	19.4458	19.4624	19.4791	19.4956
3	10.1280	9.5521	9.2766	9.1172	9.0135	8.9406	8.8452	8.7855	8.6602	8.6166	8.5720	8.5267
4	7.7086	6.9443	6.5914	6.3882	6.2561	6.1631	6.0410	5.9644	5.8025	5.7459	5.6877	5.6284
5	6.6079	5.7861	5.4095	5.1922	5.0503	4.9503	4.8183	4.7351	4.5581	4.4957	4.4314	4.3654
6	5.9874	5.1433	4.7571	4.5337	4.3874	4.2839	4.1468	4.0600	3.8742	3.8082	3.7398	3.6693
7	5.5914	4.7374	4.3468	4.1203	3.9715	3.8660	3.7257	3.6365	3.4445	3.3758	3.3043	3.2302
8	5.3177	4.4590	4.0662	3.8379	3.6875	3.5806	3.4381	3.3472	3.1503	3.0794	3.0053	2.9281
9	5.1174	4.2565	3.8625	3.6331	3.4817	3.3738	3.2296	3.1373	2.9365	2.8637	2.7872	2.7072
10	4.9646	4.1028	3.7083	3.4780	3.3258	3.2172	3.0717	2.9782	2.7740	2.6996	2.6211	2.5384
11	4.8443	3.9823	3.5874	3.3567	3.2039	3.0946	2.9480	2.8536	2.6464	2.5705	2.4901	2.4050
12	4.7472	3.8853	3.4903	3.2592	3.1059	2.9961	2.8486	2.7534	2.5436	2.4663	2.3842	2.2967
13	4.6672	3.8056	3.4105	3.1791	3.0254	2.9153	2.7669	2.6710	2.4589	2.3803	2.2966	2.2070
14	4.6001	3.7389	3.3439	3.1122	2.9582	2.8477	2.6987	2.6022	2.3879	2.3082	2.2229	2.1313
15	4.5431	3.6823	3.2874	3.0556	2.9013	2.7905	2.6408	2.5437	2.3275	2.2468	2.1601	2.0664
16	4.4940	3.6337	3.2389	3.0069	2.8524	2.7413	2.5911	2.4935	2.2756	2.1938	2.1058	2.0102
17	4.4513	3.5915	3.1968	2.9647	2.8100	2.6987	2.5480	2.4499	2.2304	2.1477	2.0584	1.9610
18	4.4139	3.5546	3.1599	2.9277	2.7729	2.6613	2.5102	2.4117	2.1906	2.1071	2.0166	1.9175
19	4.3807	3.5219	3.1274	2.8951	2.7401	2.6283	2.4768	2.3779	2.1555	2.0712	1.9795	1.8787
20	4.3512	3.4928	3.0984	2.8661	2.7109	2.5990	2.4471	2.3479	2.1242	2.0391	1.9464	1.8438
30	4.1709	3.3158	2.9223	2.6896	2.5336	2.4205	2.2662	2.1646	1.9317	1.8409	1.7396	1.6230
40	4.0847	3.2317	2.8387	2.6060	2.4495	2.3359	2.1802	2.0772	1.8389	1.7444	1.6373	1.5098
60	4.0012	3.1504	2.7581	2.5252	2.3683	2.2541	2.0970	1.9926	1.7480	1.6491	1.5343	1.3903
120	3.9201	3.0718	2.6802	2.4472	2.2899	2.1750	2.0164	1.9105	1.6587	1.5543	1.4290	1.2553
∞	3.8424	2.9966	2.6058	2.3728	2.2150	2.0995	1.9393	1.8316	1.5716	1.4602	1.3194	1.0334

4b. Table of the $\mathcal{F}$ Distribution with $\alpha = .01$

Critical Values of F Test when $\alpha = .01$

$\nu_2 \backslash \nu_1$	1	2	3	4	5	6	8	10	20	30	60	∞
1	4052	4999	5403	5624	5763	5858	5981	6055	6208	6260	6313	6365
2	98.5025	99	99.1662	99.2494	99.2993	99.3326	99.3742	99.3992	99.4492	99.4658	99.4825	99.4991
3	34.1162	30.8165	29.4567	28.7099	28.2371	27.9107	27.4892	27.2287	26.6898	26.5045	26.3164	26.1263
4	21.1977	18.0000	16.6944	15.9770	15.5219	15.2069	14.7989	14.5459	14.0196	13.8377	13.6522	13.4642
5	16.2582	13.2739	12.0600	11.3919	10.9670	10.6723	10.2893	10.0510	9.5526	9.3793	9.2020	9.0215
6	13.7450	10.9248	9.7795	9.1483	8.7459	8.4661	8.1017	7.8741	7.3958	7.2285	7.0567	6.8811
7	12.2464	9.5466	8.4513	7.8466	7.4604	7.1914	6.8400	6.6201	6.1554	5.9920	5.8236	5.6506
8	11.2586	8.6491	7.5910	7.0061	6.6318	6.3707	6.0289	5.8143	5.3591	5.1981	5.0316	4.8599
9	10.5614	8.0215	6.9919	6.4221	6.0569	5.8018	5.4671	5.2565	4.8080	4.6486	4.4831	4.3118
10	10.0443	7.5594	6.5523	5.9943	5.6363	5.3858	5.0567	4.8491	4.4054	4.2469	4.0819	3.9111
11	9.6460	7.2057	6.2167	5.6683	5.3160	5.0692	4.7445	4.5393	4.0990	3.9411	3.7761	3.6062
12	9.3302	6.9266	5.9525	5.4120	5.0643	4.8206	4.4994	4.2961	3.8584	3.7008	3.5355	3.3648
13	9.0738	6.7010	5.7394	5.2053	4.8616	4.6204	4.3021	4.1003	3.6646	3.5070	3.3413	3.1695
14	8.8616	6.5149	5.5639	5.0354	4.6950	4.4558	4.1399	3.9394	3.5052	3.3476	3.1813	3.0080
15	8.6831	6.3589	5.4170	4.8932	4.5556	4.3183	4.0045	3.8049	3.3719	3.2141	3.0471	2.8723
16	8.5310	6.2262	5.2922	4.7726	4.4374	4.2016	3.8896	3.6909	3.2587	3.1007	2.9330	2.7565
17	8.3997	6.1121	5.1850	4.6690	4.3359	4.1015	3.7910	3.5931	3.1615	3.0032	2.8348	2.6565
18	8.2854	6.0129	5.0919	4.5790	4.2479	4.0146	3.7054	3.5082	3.0771	2.9185	2.7493	2.5692
19	8.1849	5.9259	5.0103	4.5003	4.1708	3.9386	3.6305	3.4338	3.0031	2.8442	2.6742	2.4923
20	8.0960	5.8489	4.9382	4.4307	4.1027	3.8714	3.5644	3.3682	2.9377	2.7785	2.6077	2.4240
30	7.5625	5.3903	4.5097	4.0179	3.6990	3.4735	3.1726	2.9791	2.5487	2.3860	2.2079	2.0079
40	7.3141	5.1785	4.3126	3.8283	3.5138	3.2910	2.9930	2.8005	2.3689	2.2034	2.0194	1.8062
60	7.0771	4.9774	4.1259	3.6490	3.3389	3.1187	2.8233	2.6318	2.1978	2.0285	1.8363	1.6023
120	6.8509	4.7865	3.9491	3.4795	3.1735	2.9559	2.6629	2.4721	2.0346	1.8600	1.6557	1.3827
∞	6.6374	4.6073	3.7836	3.3210	3.0191	2.8038	2.5130	2.3227	1.8801	1.6983	1.4752	1.0476

Index

WILEY SERIES IN PROBABILITY AND STATISTICS
ESTABLISHED BY WALTER A. SHEWHART AND SAMUEL S. WILKS

Editors
Vic Barnett, Noel A. C. Cressie, Nicholas I. Fisher,
Iain M. Johnstone, J. B. Kadane, David G. Kendall, David W. Scott,
Bernard W. Silverman, Adrian F. M. Smith, Jozef L. Teugels;
Ralph A. Bradley, Emeritus, J. Stuart Hunter, Emeritus

Probability and Statistics Section

*ANDERSON · The Statistical Analysis of Time Series
ARNOLD, BALAKRISHNAN, and NAGARAJA · A First Course in Order Statistics
ARNOLD, BALAKRISHNAN, and NAGARAJA · Records
BACCELLI, COHEN, OLSDER, and QUADRAT · Synchronization and Linearity:
  An Algebra for Discrete Event Systems
BASILEVSKY · Statistical Factor Analysis and Related Methods: Theory and
  Applications
BERNARDO and SMITH · Bayesian Statistical Concepts and Theory
BILLINGSLEY · Convergence of Probability Measures, *Second Edition*
BOROVKOV · Asymptotic Methods in Queuing Theory
BOROVKOV · Ergodicity and Stability of Stochastic Processes
BRANDT, FRANKEN, and LISEK · Stationary Stochastic Models
CAINES · Linear Stochastic Systems
CAIROLI and DALANG · Sequential Stochastic Optimization
CONSTANTINE · Combinatorial Theory and Statistical Design
COOK · Regression Graphics
COVER and THOMAS · Elements of Information Theory
CSÖRGŐ and HORVÁTH · Weighted Approximations in Probability Statistics
CSÖRGŐ and HORVÁTH · Limit Theorems in Change Point Analysis
DETTE and STUDDEN · The Theory of Canonical Moments with Applications in
  Statistics, Probability, and Analysis
DEY and MUKERJEE · Fractional Factorial Plans
*DOOB · Stochastic Processes
DRYDEN and MARDIA · Statistical Analysis of Shape
DUPUIS and ELLIS · A Weak Convergence Approach to the Theory of Large Deviations
ETHIER and KURTZ · Markov Processes: Characterization and Convergence
FELLER · An Introduction to Probability Theory and Its Applications, Volume 1,
  Third Edition, Revised; Volume II, *Second Edition*
FULLER · Introduction to Statistical Time Series, *Second Edition*
FULLER · Measurement Error Models
GHOSH, MUKHOPADHYAY, and SEN · Sequential Estimation
GIFI · Nonlinear Multivariate Analysis
GUTTORP · Statistical Inference for Branching Processes
HALL · Introduction to the Theory of Coverage Processes
HAMPEL · Robust Statistics: The Approach Based on Influence Functions
HANNAN and DEISTLER · The Statistical Theory of Linear Systems
HUBER · Robust Statistics
IMAN and CONOVER · A Modern Approach to Statistics
JUREK and MASON · Operator-Limit Distributions in Probability Theory
KASS and VOS · Geometrical Foundations of Asymptotic Inference

*Now available in a lower priced paperback edition in the Wiley Classics Library.

*Now available in a lower priced paperback edition in the Wiley Classics Library.

*Now available in a lower priced paperback edition in the Wiley Classics Library.

*Now available in a lower priced paperback edition in the Wiley Classics Library.

*Now available in a lower priced paperback edition in the Wiley Classics Library.

Texts and References Section

*Now available in a lower priced paperback edition in the Wiley Classics Library.

Texts and References (Continued)

KOTZ and JOHNSON (editors) · Encyclopedia of Statistical Sciences: Supplement Volume

KOTZ, REED, and BANKS (editors) · Encyclopedia of Statistical Sciences: Update Volume 1

KOTZ, REED, and BANKS (editors) · Encyclopedia of Statistical Sciences: Update Volume 2

LAMPERTI · Probability: A Survey of the Mathematical Theory, *Second Edition*

LARSON · Introduction to Probability Theory and Statistical Inference, *Third Edition*

LE · Applied Categorical Data Analysis

LE · Applied Survival Analysis

MALLOWS · Design, Data, and Analysis by Some Friends of Cuthbert Daniel

MARDIA · The Art of Statistical Science: A Tribute to G. S. Watson

MASON, GUNST, and HESS · Statistical Design and Analysis of Experiments with Applications to Engineering and Science

MURRAY · X-STAT 2.0 Statistical Experimentation, Design Data Analysis, and Nonlinear Optimization

PURI, VILAPLANA, and WERTZ · New Perspectives in Theoretical and Applied Statistics

RENCHER · Methods of Multivariate Analysis

RENCHER · Multivariate Statistical Inference with Applications

ROSS · Introduction to Probability and Statistics for Engineers and Scientists

ROHATGI · An Introduction to Probability Theory and Mathematical Statistics

RYAN · Modern Regression Methods

SCHOTT · Matrix Analysis for Statistics

SEARLE · Matrix Algebra Useful for Statistics

STYAN · The Collected Papers of T. W. Anderson: 1943–1985

TIERNEY · LISP-STAT: An Object-Oriented Environment for Statistical Computing and Dynamic Graphics

WONNACOTT and WONNACOTT · Econometrics, *Second Edition*

WILEY SERIES IN PROBABILITY AND STATISTICS
ESTABLISHED BY WALTER A. SHEWHART AND SAMUEL S. WILKS

Editors
Robert M. Groves, Graham Kalton, J. N. K. Rao, Norbert Schwarz, Christopher Skinner

Survey Methodology Section

BIEMER, GROVES, LYBERG, MATHIOWETZ, and SUDMAN · Measurement Errors in Surveys

COCHRAN · Sampling Techniques, *Third Edition*

COUPER, BAKER, BETHLEHEM, CLARK, MARTIN, NICHOLLS, and O'REILLY (editors) · Computer Assisted Survey Information Collection

COX, BINDER, CHINNAPPA, CHRISTIANSON, COLLEDGE, and KOTT (editors) · Business Survey Methods

*DEMING · Sample Design in Business Research

DILLMAN · Mail and Telephone Surveys: The Total Design Method

*Now available in a lower priced paperback edition in the Wiley Classics Library.